About the volume:

The central proposition of this book is that the great anatomists of the Renaissance, from Vesalius to Fabricius and Harvey – the forebears of modern scientific biology and medicine – consciously resurrected not merely the methods but also the research projects of Aristotle and other Ancients. The Moderns' choice of topics and subjects, their aims, their evaluation of their investigations were all made in a spirit of emulation, not rejection, of their distant predecessors. This is a new perspective on the history of the 'scientific renaissance' – a history not of things found, but of projects of enquiry. It will provoke a radical reappraisal of the intellectual roots of the Renaissance as well as illuminating the current debate on the history of the body and its images.

About the author:

Andrew Cunningham is Wellcome Trust Lecturer in the History of Medicine at Cambridge University.

The Anatomical Renaissance

For Yoko

The Anatomical Renaissance

The Resurrection of the Anatomical Projects of the Ancients

ANDREW CUNNINGHAM

SCOLAR PRESS

Published by
SCOLAR PRESS
Gower House
Croft Road
Aldershot
Hants GU11 3HR
England

Ashgate Publishing Company
Old Post Road
Brookfield
Vermont 05036–9704
USA

British Library Cataloguing in Publication Data

Cunningham, Andrew
The Anatomical Renaissance: The Resurrection of the Anatomical Projects of the Ancients
1. Human anatomy—History.
I. Title.
611'.009

ISBN 1–85928–338–1

Library of Congress Cataloging-in-Publication Data

Cunningham, Andrew
The anatomical renaissance: the resurrection of the anatomical projects of the Ancients/Andrew Cunningham.
p. cm.
ISBN 1–85728–338–1
1. Human anatomy—History—16th century. I. Title.
QM11.C86 1997
611'.009'031—dc21 96–47303
CIP

ISBN 1 85928 338 1

This book is printed on acid free paper.

Typeset in Sabon by Manton Typesetters, 5-7 Eastfield Road, Louth, Lincolnshire, LN11 7AJ and printed in Great Britain by The Ipswich Book Company Ltd, Suffolk.

Contents

List of Illustrations

Credits

Acknowledgement is made to the following for permission to reproduce images in their possession:

The Wellcome Trustee: from The Wellcome Institute for the History of Medicine, London, 3.1, 3.6, 4.3–8 (the Stirling-Maxwell facsimile), 4.13, 4.20, 8.5–6; from the Wellcome Museum at the Science Museum, London: 3.5; The Syndics of Cambridge University Library, 3.2, 4.14, 5.3–4, 6.2–3; Ashmolean Museum, Oxford, 5.1; Austrian National Library, Vienna, 4.1–2; *Basler Zeitung*, Basle, 4.11; the Bibliothèque Royale de Belgique, Brussels, 4.9; the Bodleian Library, Oxford, 8.7; the George C. Izenour Archive, 3.4; National Swedish Art Museum, 4.18; the Victoria and Albert Museum, 3.3; the Warburg Institute Library, London, 4.21–2; the Whipple Library, Cambridge University, 2.2; Heinke-Brüggemann, Ch., Leipzig, 8.1; the British Library, 2.3.

Every effort has been made to trace all the copyright holders, but if any have been inadvertently overlooked the publishers will be pleased to make the necessary arrangement at the first opportunity.

Acknowledgements

I am grateful to Roger French for discussions over the years on our shared passions of the history of anatomy and of Natural Philosophy. Valuable comments have been made on the project by Adrian Wilson, Perry Williams and Don Bates. I thank Ole Grell for assisting publication. Christine Salazar has made some excellent translations from the Greek for my use, and Elisabeth Leedham-Green has produced a splendid English version of a Latin poem. The facilities of the library of the Wellcome Institute for the History of Medicine and of the Cambridge University Library have made the work possible and have been much appreciated. The work has been supported by the Wellcome Trust, to whom I offer my thanks.

Preface

There are many possible histories of anatomy. Almost all of those which currently exist are histories of discoveries, discoveries of parts and processes; they chronicle who discovered what part, in what order, when and how. The history of anatomy I offer here is different. It is a history of projects of inquiry: what approaches different people at different times took to investigating anatomy, and why. Underlying histories of discoveries is the assumption that there is somewhere out there in Nature the true body waiting to be found, and the various stages of its being found are what constitute the history of anatomy; such a history is one long story covering tens of centuries. Underlying my approach by contrast is the assumption that there is no one true body out there waiting to be found; instead I assume that there are as many true bodies as there are differing projects of inquiry to find the 'true one'. The history as I tell it is discontinuous; it is the story of different people seeking and finding different things because they are living and active at different – and particular – times and places.

In histories of discoveries it is taken for granted that inquiring into Nature is a self-evidently good, desirable and obvious thing to do. I by contrast try to treat the inquiry into Nature as being a very strange thing indeed for anyone to undertake. What needs investigation, it seems to me, is why people ever engage in such inquiry at all, since most people in most cultures neither engage in it nor care about it, and some cultures do not even have a concept of 'Nature' as something separate from themselves which they might investigate. I also believe that 'the pursuit of knowledge for its own sake' is not an adequate answer to the question of why anyone should investigate Nature, since such investigation is always culturally and historically contingent, occurring only in certain specific places and at certain specific times. So my central questions are these: Why do people investigate Nature at all? Why in the particular ways they do? Why in the particular times and places that they do? Why do they see the particular things they see? And last but not least, why these particular people?

The following essay is a first extended attempt to apply these questions to the practice of anatomy, concentrating on the sixteenth century. I see anatomizing as perhaps the oddest of all investigation into Nature on three grounds. First, because in the form in which we know it – the systematic cutting-up of dead bodies – anatomizing is unique to the western cultural tradition. Second, because of its peculiarly self-reflexive nature and the repulsion/fascination which therefore goes with it, especially within the context of a religion which treats of an immortal soul and the resurrection of the body. Third, because anatomy has the peculiarity that as a discipline and enterprise it is defined by the odd assumption (*anatomia* = cutting-up) that the functioning whole body which comes into existence as a whole and can only persist in existence as a whole, can nevertheless only properly be understood by cutting it up artificially into its supposed parts.

The present work is one of several in which, together with certain colleagues, I am investigating the identity and practice of a now obsolete discipline: Natural Philosophy.[1] This is an area of historical study which has not been attempted before. Natural Philosophy was a discipline different from science (for science was not a discipline of the early modern period), one which persisted in practice from the early thirteenth century to about 1800 AD, and anatomy was an important branch of it.

In writing the history of medicine it is very easy to forget the viewpoint of the patient. In writing the history of anatomy it is very easy to forget the viewpoint of the corpse. Only one instance is recorded of the corpse speaking – in 1538 in Lyons, as it was dissected at the hands of the famous François Rabelais. But it was recorded by Etienne Dolet, a friend of the anatomist, and thus even dissected it was heard to speak in Latin, the language of the anatomists, and to deliver what we shall learn to be the central message of the anatomists.[2] The corpse rejoices at its fate in being anatomized as 'the masterpiece of life's Great Artisan'.

**Stephani Doleti
Carmen**

Cuiusdem Epitaphium
qui exemplo edito strangulatus,
publico postea spectaculo Lugduni sectus est,
Francisco Rabelaeso medico doctissimo fabricam
corporis interpretante.

Stygem paludem, et quicquid ater Orcus habet,
  Fortuna jurarat furens,
Damnum mihi omne se reperturam, et probrum.
  Quod dum studet, atque nititur,
Et viribus totis suum in me dirum odium
  Explet, saturatque, ut collibitum est:
Tibi ecce, puncto temporis claudor carcere:
  Educor, et laqueo miser
Mox strangulatus pendeo. At quid non potest
  Fati imperium contrarii?
Ut temporis puncto perieram turpiter,
  Pari celeritate assequor,
Quod vix liceat cuiquam a Jove summo poscere,
  Spectaculo lato expositus
Secor: Medicus doctissimus planum facit,
  Quam pulchre, et affabre, ordineque
Fabricata corpus est hominis rerum Parens.
  Sectum frequens circunspicit
Corona, miraturque molem corporis
  Tanto artificio conditi.
Quare quid agat potius, quam ad extensum laqueum
  Fortuna frendens convolet,
Et morte finem odio semel quaerat suo,
  Non invidia diutius
Rumpenda? Totus ad extremum cumulor
  Honoribus, circunfluoque
Jam gloria, quem Monstrum atrox voluit rapidis
  Corvis cibum esse, et flantibus
Ludibrium ventis. Furat Sors, jam furat:
  Honoribus circunfluo.

A Poem
by Etienne Dolet

An epitaph for someone who,
having been hanged as an example to others,
was afterwards dissected at Lyons for a public spectacle,
with François Rabelais, most learned physician,
interpreting the fabric of the body.

By sludgy Styx and blackest Hell malignant Fortune swore
  She would destroy me utterly, aye, and disgrace me more;
At this she worked, for this she strove, her cup of hate to fill;
  And filled it was, and overfilled: she wrought her deadly will.
Then see me into prison thrown and, in the self-same hour
  Brought forth and hanged. Ah, what can baulk malicious Fortune's power?
And, as within a single hour my shameful doom was wrought
  In no less time I gain what may by few from Jove be sought:
Anatomised, before the crowds I lie exposed to gaze
  The while a learned doctor the Creator's art displays,
The beauty, order, workmanship by which He fashions man;
  And great the throngs that gather round the human frame to scan
And wonder at the masterpiece of life's Great Artisan.

Then what better course for Fortune, before she bursts with spite
  Than, gnashing of her teeth, to seek in agitated flight
The noose where she may end at once her being and her hate?
  While I am heaped with honours and bask in glorious state
Whom she had marked as food for crows, of playful winds the toy.
  Now Fortune you may rage indeed: all blessings I enjoy.

Notes

1. The first of these studies to appear is Roger French and Andrew Cunningham, *Before Science: The Invention of the Friars' Natural Philosophy*, Leicester, Scolar Press, 1996. For further historiographical discussion of the issues, see Andrew Cunningham and Perry Williams, 'De-centring the 'Big Picture': *The Origins of Modern Science* and The Modern Origins of Science', *British Journal for the History of Science*, 1993, 26, pp. 407–32.
2. The Latin text was first published in Book 14, no. 18, of Etienne Dolet, *Carmina Libri Quatuor*, Lyons, 1539, and is reprinted in A.-F. Le Double, *Rabelais: Anatomiste et Physiologiste*, Paris, Leroux, 1899, pp. 14–15, with a French translation. I am most grateful to Dr Elisabeth Leedham-Green for translating this into English verse at my request.

Part I
The Anatomical Renaissance

Originality means to return to the origin.

Antoni Gaudí

Introduction: The Anatomical Renaissance

It is commonly accepted that in western Europe, in the period from around 1450 to around 1600, there was a 'Renaissance': a great intellectual movement of rebirth ('re-naissance') aimed at restoring the ways of thinking and acting which had been typical of ancient Greece and Rome. This Renaissance is generally taken as marking the end of the medieval world and hence the beginnings of our 'modern' world. This movement has left its most obvious monuments in painting and architecture, but it affected music, modes of writing, philosophy and most other areas of intellectual endeavour too.

It is the common view of historians of medicine and anatomy that in the period of the Renaissance anatomy and anatomizing too underwent a rebirth; that is to say, that anatomizing became reformed according to ancient Greek principles – and that this was the great turning-point in the history of anatomy which marks the beginning of recognizably modern understandings of the functioning and anatomy of the human body. We may refer to this as 'the anatomical Renaissance'.[1]

The general interpretation offered by historians of the last 50 years about the nature of the anatomical Renaissance may be characterized as follows:

1. That Andreas Vesalius was the central figure whose efforts most contributed to effecting an anatomical Renaissance, and that his book of 1543 on the structure of the human body, usually referred to as the Fabrica, was the anatomical Renaissance put into print.
2. That this anatomical Renaissance consisted essentially of a rejection of the 'dead hand' of the ancient anatomical authorities, Aristotle and especially Galen, on the minds of Renaissance anatomists.
3. That the Ancients' doctrines were replaced by the revival of an alternative ancient Greek tradition in anatomy, that of 'seeing-for-oneself' (*autopsia*) and personal experience of dissection.
4. That the adoption of this revived tradition of investigation brought into being the beginnings of modern, scientific, anatomy.[2]

We can take Charles O'Malley, the most recent biographer of Vesalius, as a representative historian holding this point of view. In the opening passage of his biography of Vesalius O'Malley states: 'Andreas Vesalius is now recognized by most scholars as the founder of modern anatomy . . .

it was Vesalius who made the foundation secure by his factual contributions and, more important, by his method of presentation and by the scientific principle he enunciated as fundamental to research. His constant reiteration of this principle obliged anatomy to remain modern . . .'[3] Writing elsewhere on Vesalius, O'Malley expounds on what this principle was which 'obliged anatomy to remain modern' (in O'Malley's view):

> More important than the anatomical information contained in the *Fabrica* was the scientific principle enunciated therein. This was beyond criticism, fundamental to anatomical research, and has remained so. It was not difficult to demonstrate Galen's errors of anatomy, but such a demonstration was only a means to an end . . . no one had proposed a consistent policy of doubting the authority of Galen or of any other recognized authority until the only true source of anatomical knowledge – dissection and observation of the human structure – had been tested. With the publication of the *Fabrica* all major investigators of anatomy were compelled to recognize the new principle, even though at first some paid no more than lip service to it.[4]

This understanding that the Vesalian moment made anatomy scientific, in the modern sense of the term 'scientific', is shared by virtually all historians of anatomy. Even a historian as sophisticated as Jerome J. Bylebyl, can head one of the chapters of his thesis 'Vesalius, Colombo, and Falloppia and the development of scientific anatomy'; he does not spell out what 'scientific anatomy' means, probably because he can assume an understanding of the term on the part of his readers. Similarly, he does not question the view that Vesalius's work was the most basic contribution to making anatomy scientific.

So according to historians such as O'Malley, in anatomy the Renaissance did not consist of reviving the actual projects or views of the ancient anatomists themselves, but of reviving those most fundamental methods of the investigation of Nature, originated by the ancient Greeks (the methods, that is, of observation and personal experience) which had not been sufficiently rigorously followed either by the Ancients themselves or by their later medieval followers. In the eyes of such historians of anatomy, the adoption of these basic methods (observation and personal experience) thus allowed the anatomists of the sixteenth century to see more clearly the truths of the body that lay in front of their eyes, without their sight any longer being obscured by having to kow-tow to the authority and the letter of the texts of the Ancients. It thus allowed them to be properly scientific in their approach to anatomy. Vesalius's generation of anatomists was the first that began to do this, so it is regarded as inevitable that they did it still imperfectly, but their work is taken to mark the beginning of modern scientific anatomy because their example encouraged later workers to see yet more clearly.

There would, I think, still be little dispute still today with this broad characterization of what constituted the Renaissance in anatomy, at least outside the small world of professional historians of anatomy and medicine. But in recent years, within that small world, historical research has indicated that the picture is somewhat more complex than this. Interest has focused, perhaps inevitably, on William Harvey and his discovery of the circulation of the blood, a discovery made in England in the 1610s and announced publicly in 1628. Harvey is seen as the crowning point of the anatomical Renaissance and as the most modern of all Renaissance anatomists in that he brought to light this fundamental phenomenon which underlies all our modern understandings of the functioning of the body and which was quite unknown to the Ancients and medievals. A number of historians, led by Walter Pagel,[5] have pointed out how frequently Harvey remarked on his indebtedness to, and admiration of, the ancient Greek philosopher Aristotle; not just his indebtedness to Aristotle's attitude toward observation and personal experience, but to Aristotle's writings on what constitutes proper knowledge and how it is to be gained, and to his particular anatomical doctrines too. Instead of rejecting Aristotle, as the general characterization of the anatomical Renaissance given above would have led us to expect, we now find instead that Harvey – the one we see as the most modern of all Renaissance anatomists – explicitly and avowedly praised Aristotle and acknowledged the high importance that Aristotle had for him and the guidance for his own work that he found in the work of Aristotle.

> I am treading in the footsteps of other people who have borne a light in front of me in this business [Harvey wrote], and I employ their terms as much as possible. Above all others I follow Aristotle from among the Ancients, Hieronymous Fabricius ab Aquapendente from among the Moderns; I follow Aristotle as Leader, Fabricius as Pathfinder.[6]

In recognizing the veneration in which Harvey held Aristotle, recent historians of anatomy have accommodated in different ways this seeming contradiction over what the anatomical Renaissance consisted of. If in making his great modern discovery Harvey admired and followed Aristotle the Ancient, does this mean that Harvey was less modern than we thought (which is basically Pagel's position)? Or does it mean (as another historian, Gweneth Whitteridge, seems to think) that while we may continue to regard Harvey as being as modern as ever, we now need to look at the Ancients, and particularly Aristotle, as having been more modern than we thought? Strictly speaking this interpretation would of course be a contradiction in terms, since thereby the Ancients become the Moderns! Either way of coping with this dilemma implies questions about the precise form and nature of the anatomical Renaissance, and about the way in which the Renaissance supposedly constituted the great turning-point which separated the ancient from the modern views of the body.

A few years ago, in pursuit of an understanding of what Harvey meant when he said he was following 'Aristotle as Leader, Fabricius as Pathfinder' in his anatomical work, I decided to look at the anatomical enterprise of Fabricius, Harvey's own teacher at Padua. I speak of Fabricius's 'anatomical enterprise' rather than of his 'anatomical achievements', because I was deliberately refraining from seeking apparent modern features in his work; instead I was trying to discover what Fabricius's general undertaking or project was, in his own terms. These are the kinds of question I was asking: How did Fabricius choose topics to explore? How did he explore them? How did he investigate his material? What was the nature of this material? What constituted an answer for him? At what point did he think a particular inquiry was finished? (or, to put it another way, how did he know when to stop?), and so on. What I discovered was that Fabricius was trying to resurrect the research programme of Aristotle in anatomy.[7] Although it conforms to what Fabricius himself says he was doing, and also to what Harvey says about Fabricius and Aristotle being his only mentors, this was nevertheless a quite novel finding in the history of anatomy. For it gave a completely new role to the anatomizing of one of the Ancients, supposedly the deadest and deadliest of the dead hands of authority: Aristotle. Not only had the anatomizing of Aristotle not (as we used to think) been rejected by the Moderns of the sixteenth century, and not only had it been praised by some of them (as we had begun to recognize), but it had actually been deliberately selected by one of them, Fabricius, to be the model and pattern for his own practice of anatomy.

As a result of this work I came to think that the anatomical Renaissance as a whole ought to be seen not as a matter of the rejection of the doctrines and practices of the Ancients who performed anatomy, but as an attempt to emulate them, and to emulate not only their methods (observation and seeing-for-oneself in dissection) but also their entire projects of investigation, as those projects were understood by the sixteenth-century anatomists. It seemed to me that this project of Fabricius differed radically from the anatomizing of Vesalius in the previous generation. So in that piece on Fabricius I suggested that there might well have been other, different, anatomical projects being carried out in the Renaissance – different from that of Fabricius – and which were attempts to revive the anatomical work of other Ancients of anatomy. The implication of this, in turn, is that there could have been as many anatomical projects revived in the Renaissance as there had been practised in Antiquity! This again was a novel thought, for up till now it has been taken for granted by historians that the history of anatomy is one single monolithic history; that, since Antiquity onwards, just one project of anatomy has been undertaken by all anatomists, to which each anatomist has made such contributions as he was able.

The present book is the product of working through these hunches and asking both of the Ancients of anatomy and of some of their Renaissance (Modern) counterparts: what project of anatomising were they engaged in? I found that they were indeed engaged in several differing projects and that the Modern projects of the sixteenth century were attempts to recover and practise anew the projects of one or other of the Ancients. The Modern projects really were attempts to give new birth (re-naissance) to the Ancient projects, not to reject them.

If it is indeed the case, as I argue in this book, that the Ancients, and the Moderns who followed them, were engaged in a number of different anatomical projects, and not just one common monolithic project, then it follows that they saw with their mind's eye a number of different bodies: that Aristotle saw and was talking about a different concept of the body to that seen and talked about by Galen, and similarly for their Modern followers in the Renaissance. For in the first place, as we shall see, not everyone was studying the human body. And among those who were, there was disagreement about what the natural boundaries within it were, what the role of particular organs was, what pathways existed within the body, and what the relation was of certain sets of organs to others.

That anatomists of the past may have seen different bodies is again a novel position to take, for we have hitherto assumed that all anatomists were looking, or trying to look, at one and the same body – the human body as we now take it to be – and that their success or lack of success in doing so can be assessed according to what we now take that human body to be truly like. Indeed it is on this basic criterion that historians have customarily made their judgements about which past anatomists deserve credit for their work, and which deserve blame for not seeing what was in front of their noses. But what we can take to be 'really there' in Nature does not depend simply on Nature but on who is doing the looking: how a given person looks largely determines what can be seen. And how you look depends, in turn, on why you came to be looking in the first place. In other words, what you see and can see in Nature depends above all on your project of inquiry. Whether or not, as a prospective inquirer into Nature, you have become self-conscious about your project of inquiry (and most people, of course, are not), it is nevertheless your project of inquiry which provides you with the intellectual spectacles through which you look, and the mental categories which enable you to make sense of what you see through them. It is impossible for us to look at Nature without intellectual spectacles of one kind or another. Different projects of inquiry – different spectacles – make different bodies visible in anatomy.

So, with the concept of different bodies appreciated, we may be able now to make more sense of what these people did see – and why they didn't, why they couldn't, see what they didn't see. That is to say, a more sophisticated understanding of the history of anatomical investigation is now available to us than the simple assumption that it is all 'obvious' and anatomical advance and discovery is simply a matter of being willing to use your eyes and gain personal experience of dissection. So I offer this not just as an account of Ancient and Renaissance anatomy, but as an example of a more profitable approach to understanding the history of anatomy and anatomizing as a whole.

In pursuing the themes of the present book I cannot claim to have produced an exhaustive account even of the anatomical practices of the sixteenth century. I have dealt with only a few moments, and I have deliberately restricted my treatment to figures already well known in the history of anatomy. I also appreciate that my approach to the history of anatomy is somewhat unusual in not being a technical history. But I hope that approaching the history of anatomy via projects of inquiry, as I do in Part I, and then opening up the question of how we might begin to seek causal explanations for why people took up such projects, as I do in Part II, will not only be valuable in itself, but will also throw new light on the traditional history of discoveries and even help us to understand more about the technical aspects of the history of anatomy. It may help us understand why particular things came to be visible when and where they did, and to the particular people who first saw them. I hope that my account, especially in Part II, will also help reorient the traditional ways of discussing the issue of the supposed relationship of science and religion in the past.

Since I began writing this book, the history of the body has been taken out of the hands of historians of anatomy and become a very fashionable subject of academic historical discourse.[8] I should say that my work stands completely outside such new traditions of writing, especially those interpretations which discount or deny the importance of intentionality on the part of historical actors. As the present volume will demonstrate, it is my conviction that until we have understood the intentionality of past actors we have not understood the nature and identity of their actions.

Notes

1. The first use of the term that I know of is by Loren C. McKinney, 'The Beginnings of Western Scientific Anatomy: New Evidence and a Revision in Interpretation of Mondeville's Role', *Medical History*, 1962, 6, pp. 233–9, see p. 233. That there is continuing value in the concept of the Renaissance in the history of medicine has been recently claimed by Vivian Nutton, 'Greek Science in the Sixteenth Century Renaissance', in J.V. Field and Frank A.J.L. James, *Renaissance and Revolution: Humanists, Scholars, Craftsmen and Natural Philosophers in Early Modern Europe*, Cambridge University Press, 1993, pp. 15–28.
2. See his 'Cardiovascular Physiology in the Sixteenth and Early Seventeenth Centuries', Unpublished Ph.D. Thesis, Yale University, 1969.
3. Charles O'Malley, *Andreas Vesalius of Brussels, 1514–1564*, Berkeley, University of California Press, 1965, p. 1.
4. See his entry on Vesalius for the *Dictionary of Scientific Biography* (1976), p. 11, col. 1.
5. Walter Pagel, *William Harvey's Biological Ideas*, New York, Karger, 1967; Gweneth Whitteridge, *William Harvey and the Circulation of the Blood*, London, Macdonald, 1971.
6. Preface of Harvey's *Exercitationes de Generatione Animalium*, London, Pulleyn, 1651, my translation.
7. See my 'Fabricius and the "Aristotle Project" in Anatomical Teaching and Research at Padua', in A. Wear, R.K. French and I.M. Lonie, eds, *The Medical Renaissance of the Sixteenth Century*, Cambridge University Press, 1985, pp. 195–222; I deal with these issues again in the present book.
8. For instance (in no particular order): Drew Leder, ed., *The Body in Medical Thought and Practice*, Dordrecht, Kluwer Academic Publishers, 1992; Lucy Gent and Nigel Llewellyn, *Renaissance Bodies: The Human Figure in English Culture c. 1540–1660*, London, Reaktion Books, 1990; Marie-Christine Pouchelle, *The Body and Surgery in the Middle Ages*, Cambridge, Polity Press, 1990 (original French edition 1983); Francis Barker, *The Tremulous Private Body: Essays on Subjection*, London, Methuen, 1985; Jan C.C. Rupp, 'Matters of Life and Death: The Social and Cultural Conditions of the Rise of Anatomical Theatres, with Special Reference to Seventeenth Century Holland', *History of Science*, 1990, 28, pp. 263–87; Michel Fehler, Ramona Naddoff, and Nadia Tazi, eds, *Fragments for a History of the Human Body*, 3 parts, New York, Zone Books, 1989; Jonathan Sawday, 'The Fate of Marsyas: Dissecting the Renaissance Body', in Lucy Gent and Nigel Llewellyn, *Renaissance Bodies: The Human Figure in English Culture c. 1540–1660*, London, Reaktion Books, 1990, pp. 100–35; Jonathan Sawday, *The Body Emblazoned: Dissection and the Human Body in Renaissance Culture*, London, Routledge, 1995.

CHAPTER ONE

The Ancients of Anatomy

Our story will not be primarily about *things found*, as is usual in histories of anatomy, but about *projects of inquiry* as undertaken by certain anatomists. Our main story will be about the recovery and reintroduction into practice by the Moderns of the sixteenth century, of what they believed to have been the anatomical projects of the Ancients. The Ancients, of course, are the people of classical Greece and Rome. It is therefore necessary for us first to have some understanding of the anatomical projects of the Ancients who engaged in anatomy or whose writings were significant for subsequent anatomizing, taking them chronologically. As explained in the Introduction, since historians of anatomy have hitherto tended to see all anatomical enterprise in the past as part of just one monolithic anatomical project stretching from Antiquity to the present, no historian has looked at the *distinctiveness* of the anatomical projects of the Ancients before. I present the different ancient anatomical projects here in a direct manner and without polemical argument against the views of other historians. I have established the characteristics of these projects by looking at what the Ancients in question were actually doing in anatomy, and taking seriously what they said they were doing, where this can be discovered.

Chronologically first among the Ancients whom we might expect to have engaged in anatomy is Hippocrates, the supposed author of the earliest and most fundamental treatises on medicine in the western tradition (some 60 or so, most of them dating from the fifth century BC). Legend had it that he had written a large book on anatomy, which was subsequently lost, and some of the surviving treatises show some evidence of anatomical knowledge. But the Hippocratic works do not contain enough anatomical material for Hippocrates ever to have been seriously looked on as the first authority on anatomy, and thus in any direct sense his influence was negligible.

(i) Plato

In second place chronologically we come to the philosopher Plato, active and teaching in Athens in the fourth century BC (*c.* 428–*c.* 347), and who though by no stretch of the imagination an anatomist, was through his philosophical writings to have a most significant effect on subsequent anatomizing. The passage in his writings most important to anatomy happens to occur in the one book by Plato that continued to be available throughout the rest of Greek and Roman Antiquity and also throughout the so-called Dark and Middle ages in the Latin West, at least in part. Hence this book – and thus what it had to say about anatomy – was regarded as absolutely central to the doctrine of Plato throughout past centuries; today by contrast, with our present way of looking at Plato as a philosopher in our own image, this book seems like a very minor work of Plato's. The book in question is the *Timaeus*, in which Plato was teaching a mystical, religious doctrine, and the pertinent passage occurs at the point where Plato is giving a fantastical account of the creation of the universe. Today we read this simply in a metaphorical way, but it was read for centuries in a more literal way.

Plato is, in the voice of Timaeus, talking about how the world was created: 'when all things were in disorder God created in each thing in relation to itself, and in all things in relation to each other, all the measures and harmonies which they could possibly receive'. God committed the creation of mortal man to his own offspring, the lesser gods,

> And they, imitating him, received from him the immortal principle of the soul; and around this they proceeded to fashion a mortal body, and made it to be the vehicle of the soul, and constructed within the body a soul of another nature which was mortal, subject to terrible and irresistible affections, – first of all, pleasure, the greatest incitement to evil; then, pain, which deters from good; also rashness and fear, two foolish counsellors; anger hard to be appeased, and hope easily led astray; – these they mingled with irrational sense and with all-daring love according to necessary laws, and so framed man.
>
> Wherefore, fearing to pollute the divine any more than was absolutely unavoidable, they gave to the mortal nature a separate habitation in another part of the body, placing the neck between them to be the isthmus and boundary, which they constructed between the head and the breast, to keep them apart. And in the breast, and in what is termed the thorax, they encased the mortal soul; and as the one part of this was superior and the other inferior they divided the cavity of the thorax into two parts, as the women's and men's apartments are divided in houses, and placed the midriff [= diaphragm] to be a wall of partition between them. That part of the inferior soul which is endowed with courage and passion and loves contention they settled nearer the head, midway between the midriff and the neck, in order that it might be under the rule of reason and might join with it in controlling and restraining the desires when they are no longer willing of their own accord to obey the word of command issuing from the citadel.
>
> The heart, the knot of the veins and the fountain of the blood which races through all the limbs, was set in the place of guard,

> that when the might of passion was aroused by reason making proclamation of any wrong assailing them from without or being perpetrated by the desires within, quickly the whole power of feeling in the body, perceiving these commands and threats, might obey and follow through every turn and alley, and thus allow the principle of the best to have the command in all of them.
>
> But the gods, foreknowing that the palpitation of the heart in the expectation of danger and the swelling and excitement of passion was caused by fire, formed and implanted as a supporter to the heart the lung, which was, in the first place, soft and bloodless, and also had within hollows like the pores of a sponge, in order that by receiving the breath and the drink, it might give coolness and the power of respiration and alleviate the heat. Wherefore they cut the air-channels leading to the lung, and placed the lung about the heart as a soft spring, that, when passion was rife within, the heart, beating against a yielding body, might be cooled and suffer less, and might thus become more ready to join with passion in the service of reason.
>
> The part of the soul which desires meats and drinks and the other things of which it has need by reason of the bodily nature, they placed between the midriff and the boundary of the navel, contriving in all this region a sort of manger for the food of the body; and there they bound it down like a wild animal which was chained up with man, and must be nourished if man was to exist. They appointed this lower creation his place here in order that he might be always feeding at the manger, and have his dwelling as far as might be from the council-chamber, making as little noise and disturbance as possible, and permitting the best part to advise quietly for the good of the whole . . .[1]

The striking things about this scheme, which we shall see to be fundamental to much later anatomizing, are these: there are *three natural regions of the body*, whose boundaries in the body are the neck and the diaphragm. There is an immortal soul, which is the true man, lodged in the highest region, the head. There is a mortal soul lodged in the trunk of the body. This soul has a higher part, with good qualities, in the upper thorax, between the neck and the diaphragm. The heart, exercising passion in the cause of reason, is the guard of the safety of the whole body; and the lung supports it physically and gives it 'coolness and the power of respiration'. The lung was placed about the heart and was given air-passages to cool it. In the lower thorax, below the diaphragm, was lodged the lower part of the mortal soul, with baser qualities, in a sort of manger for this lowest soul, which is 'like a wild animal' which needs to be restrained. The three regions of the human body thus correspond to three souls: an immortal soul, and a mortal soul with two distinct parts.

As far as we know, the conceptual dividing of the body into these three regions, with these particular boundaries, was first made by Plato.

As his other writings indicate, Plato was obsessed with investigating the soul, and not at all interested in investigating the body. So although this account of human anatomy is obviously built on common knowledge of the inside of the human body, it does not seem to come from an extensive personal experience of dissection on Plato's part, and in that sense it was not a result of Nature telling Plato where the natural divisions and boundaries were. Rather, it was conceived by Plato *as an account of soul*; and thus it was drawn up to fit Plato's concept of what the soul is and how it works. Plato's version of philosophy (much of it expressed through the mouth of Socrates in Plato's dialogues) may be said to have been entirely about the soul – the immortal soul – about the soul as the true man, about the fulfilment, career and fate of the soul; about how the soul could be perfected, how it could grasp absolute concepts, 'universals' or 'Forms' as they were called, such as The Just, The Good, The Beautiful, and realize them in itself, – and hence about how the philosopher could live the Good Life, the life of the good man. Taken as a whole, Plato's philosophy was, in the words of a celebrated modern scholar, 'the morality of aspiration', and it 'implies a constant effort of the soul towards an ideal of perfection'.[2]

In his philosophy, Plato particularly wanted accounts of things which were in terms of 'the best'; in terms of intention, design, the fitness of things for their roles – categories which presuppose the operations of some Intelligence. This is very much evident also in his account in *Timaeus* of the origin and structure of the human body. For it is intrinsic to this account that the body is shown to be demarcated by Nature into its three regions *deliberately*, as the result of intention: the body is the product of intelligent design, the intelligent design of God the Creator. Each region of the body, each part of the body, is where it is and how it is for the sake of serving the soul. It is thus a 'teleological' account, and the *telos* – the goal – of the body and each of its regions, is the service of the soul. 'For the sake of' what does the human body as a whole exist, in Plato's account? For the sake of the soul.

(ii) Aristotle

The next Ancient of anatomy, and this time one who actively engaged in anatomizing, also lived in the Athens of the mid-fourth century BC, and it is the philosopher Aristotle. Pupil of Plato, grand-pupil of Socrates, Aristotle worked on and taught what seems to the modern eye to be a wide variety of topics. A large proportion of Aristotle's writings (or reports of his teachings) survive. Among them are three large and two small treatises obviously on animals. We can no longer be perfectly certain what, if any, titles Aristotle himself gave to these, but they came to be known (and are still known to scholars today) by Latin titles:[3]

1. *Historia animalium*: 'History of animals'
2. *De partibus animalium*: 'On the parts of animals'
3. *De generatione animalium*: 'On the generation of animals'

The smaller works with animals in their traditional titles are:

4. *De incessu animalium*: 'On the progression of animals'
5. *De motu animalium*: 'On the (local) motion of animals'.

It is evident from these that Aristotle had detailed knowledge of the anatomy of a wide range of animals, knowledge he seems to have acquired from his own first-hand experience of dissection, and possibly of vivisection too. He also had much information about the anatomy and habits of yet other animals, which he acquired from fishermen and others. His first- and second-hand knowledge about animals was enormous and without precedent. Scholars believe that Aristotle was the first of the Greeks to explore animal anatomy systematically and to any significant extent.

Living in a scientific age as they do, the spontaneous act of modern historians has, unfortunately, been to reclassify Aristotle's anatomical work into categories which are appropriate for today but happen to be totally inappropriate for Aristotle's time, categories such as 'biology', 'comparative anatomy' and 'embryology'; and then they have asked about Aristotle's supposed 'contribution' to the development of these sciences – none of which he was or could have been practising, since they all date from the end of the eighteenth century or the beginning of the nineteenth century AD. But the category under which Aristotle actually did his anatomy was 'philosophy' as he understood that quintessentially Greek enterprise *philo-sophia*, love of wisdom. And for Aristotle anatomy was completely integral to that enterprise of philosophy, not simply a minor section of it. In this he was like Plato. Indeed, nothing can better illustrate the gulf between our present-day view of anatomy and Aristotle's philosophical view of anatomy than the fact that Aristotle's anatomical project takes its beginnings from his concern with the soul. Any anatomical project we might engage in today would start anywhere but there.[4]

Aristotle's work on the soul was part of his attempt to reformulate the version of Greek philosophy that his teacher Plato had been so concerned to create.[5] His quarrel with it started, of course, outside the arena of philosophizing, as philosophers' differences usually do. But the soul had been made central to all philosophizing by Plato. Plato had argued that the soul plays a very important role because it is a thing of a certain kind;

Aristotle was claiming that it plays an equally important but different role because it is a thing of a different kind. As we shall see, this difference about the nature of the *soul* was to lead Aristotle to see with different eyes the nature of the *body* and its relation to that soul.

The purpose of Aristotle's book *De Anima*, 'On the soul' is to set up Aristotle's new criteria for what the soul is and can do, and how we can find out about it, and Aristotle's intention in it was to begin to undermine Plato's view of philosophy and replace it with a better one. Hence *De Anima* puts forward a new view of the role, function and nature of the soul.

One of the things that Aristotle disagreed with Plato about was whether *this* world was of interest and consequence to the true philosopher. Aristotle agreed with Plato that the perfect, the eternal and the stable were the true objects of the philosopher's quest. But Plato had claimed that such things were not perceivable in this imperfect and transitory world, only in the world of Forms; for Plato 'the world of perfect Forms contains all that is truly real'.[6] Aristotle, by contrast, claimed that the perfect, eternal and stable could indeed be seen within the imperfect and transitory things of this world.[7] As has been often remarked, this difference between them is nicely captured in the famous 'School of Athens' fresco in the Vatican painted by Raphael in 1509–12, which portrays Plato pointing upwards to the perfect, unchanging heavens and the permanent 'Forms', and Aristotle pointing downwards to the things of the changing sub-lunar world, each thereby indicating that *this* is what the true philosopher should be concerning himself with.

Aristotle's *De Anima* opens thus:

> We assume that knowledge is something beautiful and valuable, but one kind more so than another, either in virtue of its accuracy, or because it is better and more wonderful; on both these counts we can put the inquiry about the soul among the first, and it appears that the knowledge of the soul contributes greatly to the complete truth, and most of all to the truth about Nature. For the soul is in a way the principle of animals (*arche zoon*). So we seek to contemplate and come to know its nature and its essence, and then its properties.[8]

The core of Aristotle's revolution here against Plato is encapsulated in the passage 'the soul is in a way the principle of animals'. For Plato had been highly concerned with the soul, but with the human soul, and for Plato soul was distinctive of humans. What Aristotle was doing was taking this concept of soul-ness and applying it to *all* things which have life. Thus Aristotle made the two conditions, that of being alive and that of possessing soul, exactly equivalent. For Aristotle, therefore, for a thing to be alive *is* for it to possess soul, and for it to possess soul *is* for it to be alive: 'We say, then, that that which has soul (*to empsychon*) is distinguished from that which has not by living.'[9] Thus to investigate animated things is to investigate the soul, and reciprocally to investigate the soul is by definition to investigate the animated. Aristotle speaks of 'The Animated' (*to empsychon*) in the same way that Plato talked of The Good, or The Just: that is to say, as a 'universal'. In practice, as we shall see, he was actually most concerned with that aspect of 'The Animated' which has the capacity of motion, so to express Aristotle's object of study I shall use the expression 'The Animal', with capitalization to indicate that it is a universal. It is significant to note that the word *animal* (i.e. possessed of *anima*, soul) was the equivalent later Latin term for *to empsychon*, 'The Animated'.

So, Aristotle too wanted to investigate the soul, just as Plato had done, but the soul in his own new definition. The soul is the Form of The Animal: it is what makes it what it is, because it is what The Animal does. The soul is that thing which distinguishes a live animated being from a corpse, and for Aristotle the soul is the *stable Form* (that is, that which gives form) to the live things in this world amid all the apparent flux and change they undergo from their generation through their growth and decay to their regeneration.

As Aristotle says in *De Anima*, 'Just as the pupil and seeing [together] constitute the eye, so the soul and the body [together] constitute The Living (*to zoon*)': the soul is the body's power exercised.[10] Aristotle thus embarked on the anatomizing of animals in order to discover the instruments (*organa*) that the soul of The Animal uses to express itself through the body, for this is, according to Aristotle's understanding of the soul, the only possible way to investigate the soul.[11] That is to say, Aristotle's object of study in investigating animals was *not* the animals themselves and the mechanism of their operations (as it would be today), but *the soul of The Animal* as it is actually expressed through the animals which actually exist in the world.

'Beginning our investigation we say then,' wrote Aristotle,

> that that which has soul is distinguished from that which has not by living. But the word living (*zen*) is used in many senses, and we say that a thing lives if any one of the following is present in it: mind, sensation, movement or rest with respect to place, besides the movement involved in nutrition, viz. contraction and growth. Hence all plants are considered to live, for they evidently have in themselves such a power and first principle by means of which they both grow and decay.[12]

This is the first and lowest aspect of soul, this 'nutritive' or 'vegetative' soul (as it came to be called): the capacity that a plant or vegetable has for *nourishing* itself and *generating* others like itself.

Animals possess this vegetative aspect of soul, but are distinguished from plants by also having *sensation*, and of the different senses, touch is the most basic.[13] Aristotle writes:

> The soul is the origin (*arche*) of these, and is defined by them, that is by nutrition, sensation, thought, and by movement . . . in some animals all of these are present, in some only some of them, and in others only one. It is this which makes the difference between animals . . . We must then inquire in each particular case, what is the soul of each: for instance of the plant, of man, and of the beast. But we must also consider why they are thus arranged in a series . . . It is clear that the account of each of these [faculties] is also the most appropriate account that can be given of the soul.[14]

This is what Aristotle's anatomizing of animals was created to do: to investigate in an empirical manner the operation of these different faculties of the soul of The Animal and the organs through which they are exercised, and to investigate the reasons why and how different particular animals participate in different degrees of soul. To this project Aristotle devoted many treatises. For

> if one is to state what each of them is – for instance, the thinking, sensitive, nutritive [faculty] – one must first investigate what *thinking* and *perceiving* are; for logically the operations and actions come before the faculties themselves. And if this is so, and if one should have examined – even before these operations and actions – what they relate to, then for the same reason one must first of all lay down definitions about, for instance, food or 'the perceived' or 'the thought'.[15]

So Aristotle's soul-investigating project involved three inter-related investigations. The first was into the nature of the things that constitute food (what is it about them that makes them able to become 'the eaten'?); the nature of the things that can be sensed (what is it about them that makes them able to become 'the sensed'?); and similarly the nature of the things that can be thought. The second part of the project involved investigating what nutrition, generation, sensation, perception, thinking and motion are; that is to say, what movements, activities or processes they are. Then, finally, there was the investigation of the operation of these activities in animals: that is, the investigation of the organs ('instruments') that are used for such activities in different animals. Only then does one understand the soul of The Animal.

So a significant proportion of Aristotle's soul-investigation required an empirical investigation of animals and their organs – that is, an anatomical investigation, a cutting-up (which is what anatomy, *anatomia*, basically means) investigation – in order to find how the various faculties of the soul are carried out in the actual versions of The Animal which exist in the world, and how and why different animals participate in the soul to different degrees. This is the project Aristotle was engaged on and in the course of which he assembled and wrote the large books *History of Animals*, *Parts of Animals* and *Generation of Animals*.[16] For instance, as Balme has shown, *History of Animals* and *Parts of Animals* are both studies of *differentiae* (differences), aimed at finding the 'causes' of animals' parts: 'Aristotle does state his purpose: "first to grasp the differentiae and attributes that belong to all animals, then to discover their causes"',[17] and as Aristotle made clear, the whole inquiry is undertaken in order to find out about the soul of The Animal. Always Aristotle was concerned with the *causes* (or 'uses') of the parts and organs, the for-the-sake-of-which the part exists and is as it is, and always he finds this in the service of the soul. In any particular animal the incidence of a given part may differ from the norm, and in such cases Aristotle always explains the particularity by reference to either the other bodily parts, or the life, activities or character of this particular animal.[18] Thus he could reveal the extent to which, and how, every animal participates in the aspects of the soul, and he could also show how and why the aspects of the soul have a different expression in some animals.

Thus Aristotle's philosophical writings on the soul can be seen to be related in the following way. (It should be noted that although the different aspects of the soul-according-to-Aristotle were often, in a shorthand manner conventionally called, for instance, 'the vegetative soul', they are still only different aspects of the *one* soul.)

GENERAL:	On the soul History of animals Parts of animals Dissections (treatise lost) Sleep and wake* Length and shortness of life* Youth and old age*
VEGETATIVE SOUL:	On food (treatise lost) On respiration* (On breath*) On the generation of animals
SENSITIVE SOUL:	Sense and sensible objects (On things heard)
MOTILE SOUL:	On the movement of animals On the progression of animals

RATIONAL SOUL: On memory and recollection*
Dreams*

*treatises later grouped as *Parva Naturalia*
Where a title is given in brackets its authenticity is regarded as dubious.

We are now in a position to appreciate properly – appreciate, that is, as part of Aristotle's soul project, and not as some paean to the study of the as-yet non-existent science of biology – the most famous passage in which Aristotle expresses directly his desire to turn philosophizing away from obsessive concern with the eternal divine bodies in the heavens, and direct it toward things in the constantly changing sub-lunar region, especially animals. It comes from *Parts of Animals* Book 1, c. 5.

> Of the beings such as are composed by Nature – the ones we call ungenerated and imperishable for all eternity, and the others we say partake of coming-to-be and passing-away – it so happens that while the first ones [i.e. the heavenly bodies] are worthy and divine, there are very few views (*theoria*) of them available for us because, if we try to investigate these beings – even those which we long to know about – that which is clear to perception is very scanty. It is easier for us to acquire knowledge about the perishable plants and animals because we live among them, since one can acquire much knowledge about each kind that exists if one cares to take enough trouble. And each of them has its charm. For even if we grasp only a little bit about the heavenly bodies, yet by the excellence of the knowledge it is more pleasurable than all the things around us (just as seeing a fleeting glimpse of our loved ones is more pleasurable than seeing many other things large and clearly), nevertheless with respect to the others [i.e. the earthly beings] since we know more and better about them they have superiority in knowledge [over what we know of heavenly things]. And furthermore, by being nearer to us and more similar in nature, they make some compensation for the philosophy about heavenly things.
>
> As we have already treated the heavenly bodies, giving our opinion about them, it remains to speak about animal nature leaving out nothing as far as possible, neither nobler nor less noble. For even with animals which are not pleasant to look at, nevertheless the Nature at work in them holds extraordinary pleasures for those who are capable of recognising the causes (*tas aitias*) and are philosophers by nature. After all, it would be unreasonable and absurd if looking at pictures of them we rejoice that we see [both the representation of them and] at the same time an art-at-work (such as the art of painting or the art of sculpture), and yet we did not love more the contemplation (*theoria*) of the things themselves composed by Nature since we are capable of discerning the causes.
>
> Therefore one must not be childishly disgusted by the study of less worthy animals, because in all the things in Nature there is something marvelous. And just as it is said that Heraclitus said to

> some visitors who wanted to meet him and stopped when they entered and saw him warming himself by the stove – he ordered them to come in and not be afraid, 'because even here there are gods', so one must go about the investigation of every animal without expressing distaste, since in all of them there is something of Nature and of The Beautiful. For the non-random, the for-something's-sake is in the works of Nature most of all; and the thing because-of-which it is composed, or has come-into-being – its purposes (*telos*) – are part of The Beautiful. And if anyone thinks that the investigation (*theoria*) of the other animals is unworthy, then he must think the same way about himself too, for it is not possible without much disgust to see the things of which the human kind is composed, such as blood, flesh, bones, veins, and other such parts. Equally, one must recognise that he who discusses any one of those parts or equipment is not speaking about the material (*hyle*) itself and not for its own sake, but for the sake of the entire form (*morphe*) – just as one discusses a house but not the bricks, mortar, timber. In the same way one must recognise that discourse (*logos*) about Nature is about the composite-thing and the entire being as a whole, but not about those things which never occur separately from those beings.
>
> . . . Now, as each of the parts of the body, like every other instrument, is for the sake of some purpose, viz. some action (*praxis*), it is evident that the body as a whole must also exist for the sake of some complex action. Just as the saw is there for sawing and not sawing for the sake of the saw, because sawing is the using of the instrument, so in the same way the body exists in some way for the sake of the soul, and the parts for the sake of those functions for which each of them has been formed.[19]

A full account of Aristotle's soul project is beyond our present scope, but enough has now been said about it for it to be possible for us to appreciate its extent and the importance of it in Aristotle's system of philosophy; the study of the soul was absolutely central to Aristotle's philosophy, and his philosophy cannot be properly understood without appreciating the centrality of the soul to it. The study of animals – the empirical study of the soul in action, through anatomy – was not therefore something marginal or tangential to Aristotle's philosophy, but lay at its very centre. Our modern-day enterprise of philosophy, by contrast, has abandoned any concern with the soul and thus it has no room for the investigation of animals – the soul in action. The study of animals has been hived off, in a soul-less form, to a new discipline, that of biology.

With respect to the history of anatomizing that we are concerned with here, what we need to note in particular about Aristotle's soul project is that for him the heart is far and away the most important organ in the animal body, for it is the centre of the most basic aspect of the animal soul – the 'vegetative soul' which rules in the whole of the trunk of the body. The heart is the first part to come into being in the animal, and the last to die. If the soul is localized anywhere, then it is in the heart. Aristotle also argues that the heart is the centre of all the other aspects of the animal soul as well: the sensitive, the motile, the rational.[20] With the heart thus treated as the centre or seat of the vegetative soul, this means that Aristotle saw the blood vessels of the body as constituting only *one* system, and based on the heart; that is to say, he made no functional distinction between arteries and veins. Nor did he attribute much in the way of function to the brain; for him, thought had no anatomical location since the mind is solely a Form, and hence there are no organs associated with it to be investigated.[21] Such attitudes were a source of great controversy after Aristotle's time, when a functional distinction between arteries and veins was made, and when the nerves were discovered, distributed from the brain.

Finally, of course, we need to note that when Aristotle studied the soul of The Animal he was indeed studying man, but only as an instance of The Animal who happened to have more aspects of soul than any other creature – for man alone had all the aspects of soul. The soul of The Animal is Aristotle's central object of study in his anatomizing, not the soul and body of man.

And this brings us to the critical anatomical part: the diaphragm. As we saw, for Plato the diaphragm is a most important boundary in the body, separating the heart and lungs from the base organs of the belly; separating, that is, the higher part of the mortal soul from the lower part, separating that part of the mortal soul 'which is endowed with courage and passion' from the part of it which needs to be 'bound down like a wild animal' chained up with man and which 'must be nourished if man was to exist'. Plato was of course talking about man and about the soul of man. But for Aristotle, by contrast, the diaphragm is of no particular consequence. For most animals do not possess a diaphragm. Aristotle's object of study was 'The Animal', and he was primarily interested in the parts which are typical of The Animal: that is, those possessed by most animals. The diaphragm is not such a part; it is possessed by only some animals, and (like any non-universal part) its presence in them needs to be understood, Aristotle says, in relation to the particular conditions of their life, or to their other bodily parts, or to the activities or the character of these particular animals. The diaphragm is something to be explained as an *exception*, as something *unusual*, in certain animals; it is not, for Aristotle, typical of The Animal. If, as a Greek philosopher in search of the soul, your concern is with man, then the diaphragm is of crucial importance to you, marking as it does one of the great natural boundaries of the human body, put there for a reason intimately associated with the nature of the soul. But if your concern is with the soul of The Animal, then the diaphragm is of minor importance.

It may be noted that for zoologists today the diaphragm 'is diagnostic of mammals'.[22] The category 'mammal' was originated only in the eighteenth century, by Linnaeus, and could not therefore have been of consequence either in Antiquity or at any time before the eighteenth century. Man is a mammal, but most animals are not.

As we shall see, we can take the diaphragm and concern with it as a 'marker' to distinguish the anatomical tradition which looked back to Plato from that which looked back to Aristotle. It is a 'marker' in two senses: concern with it indicates a concern with the Platonic view of the body, and it is seen as a physical marker which marks off one set of internal organs from another, both functionally and conceptually. Together with the differing doctrines of Plato and Aristotle on the importance of the heart and what kind of organ it is, on the functions of the brain (and nerves), and on the respective contributions of the male and female in generation, the diaphragm is one of the features which most distinguishes the 'different bodies' that Plato and Aristotle saw: respectively, the human body, divinely planned, and The Animal, teleologically arranged.

(iii) Herophilus and Erasistratus

The next Ancients of anatomy, according to the information that has survived, were two characters active, it is generally believed, in Alexandria in about the third century BC: Herophilus and Erasistratus. The city of Alexandria in Egypt was one of several Alexandrias founded and immodestly named after himself by Aristotle's most famous pupil, Alexander the Great, in the course of his tour of conquest of the Persian empire. On Alexander's death the spoils of his empire were divided among his generals, and Ptolemy got Egypt. For something like a thousand years (*c.* 300 BC to *c.* 700 AD) Alexandria remained an oasis of Greek ('Hellenistic') culture, and the Ptolemies founded in it the famous Museum (literally 'Home of the Muses') and Library. It is believed, on no particular evidence, that Herophilus and Erasistratus had some connection with these institutions. None of the writings of either of them survive, though information about their opinions is available to us from the writings of Galen (whom we shall discuss in a moment).

By the period in which Herophilus and Erasistratus lived, anatomists had made two crucial distinctions which had not been known to Aristotle. First they had distinguished between two kinds of vessel running to and from the heart: the arteries and the veins. The arteries were so-called on the grounds that they contained air (*aer* is the linguistic root of *arteria*), and were hence seen as respiratory vessels, the vessels which transmitted vitality, life, to the body. The veins, by contrast, were the blood vessels based on the liver and which transmitted nourishment to the body after the food ingested had been digested by the stomach. Once this functional distinction had been made, then other marks differentiating the functional roles of arteries and veins became evident: only arteries pulsate, arteries have thicker walls, the blood they contain (if you believed they do indeed contain blood) is of a brighter colour, death supervenes if an artery is cut since the life-giving *pneuma* escapes, whereas someone can easily survive the cutting of a vein, since only blood-as-food is lost. Similarly, the nerves had been distinguished from what Aristotle had called the 'tendons', and their origin in the brain had been traced.

Herophilus and Erasistratus may have had some role in these discoveries since they all depend on anatomical inspection of the living creature and Herophilus and Erasistratus became particularly famous later for having *vivisected men* in the course of their anatomical investigations. What better way to discover the existence and function of nerves, and to trace their paths? What better way to see that the vessels which run to and from the heart and the liver have different behaviours, contain different things (air, blood), necessarily have different functions?

Whether Herophilus and Erasistratus had in fact engaged in the vivisection of men is, for our purposes, neither here nor there. But what does matter for our purposes is that there was a report that they had done so, in a Roman work of the first century AD: the *De medicina* ('On medicine') of Celsus. However, this report was unknown to later anatomists until it was discovered in the mid-fifthteenth century, and printed in 1478 at Florence.

This is the passage. Celsus is saying that the Dogmatist (or Rationalist) sect of physicians claim that, since pains and various kinds of diseases are generated in the internal parts of the body, no one can properly apply remedies for these unless he is familiar with the internal parts. Hence the Dogmatists cut open the bodies of the dead to scrutinize their viscera and intestines: and the Dogmatists say

> that Herophilus and Erasistratus did this in the best way by far: they dissected living criminals received out of prison from the kings, and they considered – with the spirit still present in the criminals – the things which nature had hitherto hidden, and their position, colour, figure, magnitude, sequence, hardness, softness, smoothness, contact, and then the protuberances and depressions of each of them . . .

> [The Dogmatists say] that it is not cruel (as most people claim) that remedies for innocent people of all later ages should be sought from the sufferings of criminals, and of only a few of them at that.[23]

This account of Herophilus's and Erasistratus's predilection for the vivisection of humans, is introduced by Celsus in the midst of a critical comparison he is making of the different medical sects of his own time: the Rationalists (or Dogmatists), the Empirics, and the Methodists.[24] Herophilus and Erasistratus are, for Celsus, in the rationalist tradition. This is the tradition which springs from Aristotle, and which takes it as axiomatic that we can gain reliable knowledge of the *function* the different parts of the body fulfil in health and disease from an investigation of their *structure* – by cutting bodies up, either dead (dissection) or alive (vivisection). It is rational because disciplined reasoning is taken to be the key for going from the seen to the unseen (which means both from the function to the structure and from the structure to the function). We can, that is, discover from anatomy-informed-by-strict-reasoning, the *causes* why the body is as it is and works the way it does both in health and disease. This is what, in Celsus's report, Herophilus and Erasistratus had been up to: looking at the position, colour, figure, magnitude, and so on of the parts normally hidden during life, in order to discover (by the use of strict reasoning) how they really are and how they work.

Whether or not Celsus's characterization of the medical sects of his own time can be taken at face value, his account does illustrate the fact that the rationalist approach is not, and historically has not been, the only possible approach to take to anatomizing, even within the Greek tradition. Celsus's Empirics, for instance, who were as Greek in their view of the world as were his Rationalists, are reported by him as maintaining that the information acquired from anatomizing or vivisecting the human body simply cannot be used to infer how the body works while it is intact and alive. We cannot be certain even that the colour of parts and organs is the same in the intact as in the dissected/vivisected body, let alone their shape or functioning. And Celsus claims that the Methodists of his day, equally Greek by cultural heritage, believed that knowledge acquired from anatomizing bore no relation whatsoever to the knowledge required by the physician for healing the body: the body works on a level quite inaccessible to the anatomist! Thus there seems to have been in Celsus's day a variety of Greek attitudes to the value of anatomizing, with some people seeing it as of no value whatsoever, while others saw it as crucial, especially for a physician. Given this, we need hardly be surprised to find varieties of opinion within the rationalist tradition itself; that it embraced ways of doing anatomy which were divided by almost as many differences between themselves, as they were united by sharing the basic rationalist presupposition that in anatomy structure is the key to function and that function is achieved by structure.

This is the case here with Herophilus and Erasistratus as compared with Aristotle, whose respective projects in anatomizing (if we can put faith in this account by Celsus from many centuries later) were both rationalist, yet which differed radically in aim and method. For where Aristotle had been investigating The Animal in order to find out about the soul of The Animal, Herophilus and Erasistratus are evidently concerned only with the anatomy of the human, and their interest in that seems to have been directed toward medicine and the finding of cures, rather than toward knowledge of the anatomy of the human body for its own sake or (as far as we know) for philosophical knowledge about the nature of the human soul.

There is much disagreement and uncertainty among modern scholars about the precise doctrines of Herophilus and Erasistratus, and since much of what is known comes from attacks on Erasistratus centuries later by Galen, we shall return to them in the next section.

(iv) Galen

The last and in many ways the most important of the Ancients of anatomy was also in the Greek rationalist tradition: it was Galen. While there were many anatomists between Herophilus and Erasistratus and Galen, their works are lost to us and are not consequential for our purposes.

By some freak of circumstance no longer accessible to us, Galen became the dominant figure in medicine of the Greek tradition from shortly after his death in about 200 AD until around 1700 AD and even beyond. In Alexandria until about 700 AD, in Byzantium from about 650 to about 1450 AD, in the Arab empire from about 850 AD (and indeed still today in modern Pakistan), in western Europe from about 1000 AD, it is no exaggeration to say that literate medicine, academic medicine, 'official' medicine *was* Galen. For the works of almost no other ancient physician apart from 'Hippocrates' have survived in any quantity and, such as do survive from later Alexandria or Byzantium, are in large part compilations in one form or another of Galen's writings. On the other hand, Galen's writings have survived in enormous bulk.

Galen himself lived and worked mainly in Rome, and within the context of the Roman empire. But in his time medicine was still largely a Greek affair, and most physicians were Greeks. Galen's language was Greek, and all his writings were in Greek. What is more, the tradition in which they are written is also totally Greek; Galen places himself, quite deliberately, in the tradition of Hippocrates, Plato and Aristotle. A Greek physician by occupation, Galen was a Greek philosopher by inclination. He welds his reading of Plato and Aristotle to his presentation of the teachings of the venerable 'father' of medicine, Hippocrates. Galen presents himself as the only authentic interpreter of Hippocratic – true and genuine – medicine, and the interpretation he presents is highly rationalistic in character and shot through with the philosophy of Plato and of Aristotle.[25] He had of course read Aristotle's books on animals (as he had read all Aristotle's other available writings), and he had taken them into account. Nevertheless, however reminiscent of Aristotle his writings may seem to us when he is talking about anatomy, we must remember that Galen's occupation was not that of a full-time philosopher, and hence what he was up to (and thus what activity his writings represent) was different from what Aristotle had been up to.

What Galen was creating was a fully coherent and integrated understanding, by a practical man for practical men, of medicine for the treatment of man. Virtually every one of his surviving writings bears on this in some way.[26] There are introductory books for the beginner advising him what medical sect to follow, what things to learn and in what order. There are major treatises on diagnostics, on the pulse, on diet, on antidotes, on the causes of disease, on compounding medicines, on hygiene, on therapeutic method, on the natural faculties, and so forth. Whether the occasion of the work was a commentary on a work of Hippocrates, or an attempt to ridicule one of his own contemporaries (in which he delighted), virtually all the things Galen wrote were ways of teaching the truth as he saw it about health, disease, and cure. His learning was prodigious and his output enormous.

Among these many books promoting one or other aspect of the system of medicine that Galen believed in, there were a number on anatomy. It was above all from Galen that later ages learnt that a knowledge of anatomy must lie at the basis of a proper knowledge of medicine. His whole medical system presumes a detailed knowledge of human anatomy, and Galen was happy to provide the books which would assist one in learning this anatomy. There are two large works of the most enormous importance; they are usually referred to by Latin titles:

1. *De anatomicis administrationibus*: 'On anatomical procedures' or 'On anatomical administrations'.
2. *De usu partium*: 'On the use of parts'.

There is also a work *On bones, for beginners*, and several treatises on the dissection of particular organs, such as the eyes, the nerves, the muscles, the vocal cords and the uterus.[27]

The sequence of learning anatomy that Galen recommended in *On anatomical procedures*, was that one should first learn about the *bones*, because they play the same role in the human body as walls do in houses. That Galen took such a view of the relation of the skeleton to the flesh and organs may owe something to his experience of the building techniques of his time – the use of load-bearing pillars and beams, the space between them being filled in with curtain walls – and to the fact that his own father was an architect by profession. He writes:

> Since, therefore, the form of the body is assimilated to the bones, and the nature of the other parts follows them, I would have you first gain an exact experience of human bones. It is not enough to study them casually or read of them only in a book: No, not even in those books that some people have entitled *Osteologia*, others *Skeletons*, and yet others simply *On bones* – as is ours, which I am convinced excels all those written before it in exactness of things (*pragmaton*), brevity and lucidity.[28]

That estimate of the quality of his own work is typically Galen. After the bones, one should proceed as follows:

> The nature of all the bones, as I said, is to be thoroughly learnt either from man, or from the body of the ape, or better from both. Then one should move on to the anatomy of the muscles. For these two parts of the body underlie all the others, like the foundations of a building. After that you can learn whatever you want to learn first, whether it be the arteries or the veins or the nerves. And for someone skilled in the anatomy of these, should come study of the nature of the viscera. And then there remains the knowledge of the intestines, fat and glands. One has to examine these by themselves in more detail. It is better for you to practise this in this order.[29]

On anatomical procedures, near the beginning of which this passage occurs, is the work that pursues the practical dissection of these parts and in the order Galen has here specified. This work had 15 'books' or sections, of which the last six and a half were unknown in the West after Galen's day until the twentieth century. The work, as known to later anatomists, went only as far as book 9, on the brain.[30] *On anatomical procedures* was intended to accompany one through dissection; but it will be noticed that the sequence Galen adopts demands that one should have more than one body available for this purpose, since the skeleton and bones are the last things one reaches in a dissection but are the starting-point for learning about anatomy in Galen's method. This matched Galen's own practical procedures, for he had available to him quite a ready supply of apes to dissect. Indeed Galen never dissected a human body, because this was not permitted in the Rome of his day (though it may still have been possible in Alexandria). Galen makes no secret of the fact that he is acquiring, and urging his followers to acquire, a detailed knowledge of *human* anatomy from the dissection of *apes*; to minimize possible false inference, he recommends that his followers use the apes which are 'most like man', today identified as Barbary apes. Moreover, he says, one should seize every chance one has of inspecting the human cadaver, and he instances the inspection of the bones of some corpse whose flesh had been washed away by the action of a stream in which it had died.

So Galen's dissection sequence began with the bones and skeleton and moved on to the muscles, 'For these two parts of the body underlie all the others, like the foundations of a building.' Thereafter one should learn the three systems of vessels which run throughout the body: the arteries, the veins, and the nerves. Finally, one should learn the anatomy of the viscera, the organs contained within the trunk of the body and the head.

Thus Galen recommends one particular order of dissecting in *On anatomical procedures*. But when it came to understanding what the different parts and organs of the body do with respect to the whole, he deliberately and explicitly followed a different sequence, the account Plato gave in the *Timaeus* of the human body being divided according to three grand functions. All of Galen's understanding of how the body functions is based on this.

This allegiance to Plato's view is most clearly expressed in the opening words of Galen's book *On the natural faculties*:

> Since sensation and voluntary motion are peculiar to animals, whilst growth and nutrition are common to plants as well, we may look on sensation and voluntary motion as the work of the *soul*, and growth and nutrition as the work of *Nature*.
>
> And if there be anyone who allows a share in soul to plants as well, and separates the two kinds of soul, naming one kind *vegetative*, and the other *sensory*, this person is not saying anything else, although his language is unusual.
>
> ... we say that animals are governed at the same time by their soul and by Nature, and plants by Nature alone, and that growth and nutrition are the effects of Nature, not of soul.[31]

Growth, nutrition (and generation too, as Galen later makes clear) are activities performed by *Nature*: they are simply the work of Nature. (Galen then argues in this work that they can each be understood as a sequence of *faculties* – faculties of attraction, retention, assimilation and expulsion of nourishment – both in animals and plants.) But, Galen claims, *sensation* and *voluntary motion* are of a quite different kind, for only animals perform them; they are a product of the activity of the soul, of the 'psychic' soul, as he calls it, located in the head, which feels and also controls voluntary movement.[32] (In Latin this psychic soul came to be called the 'animal' soul.) Thus the nutritive, growth and generative functions, which Plato had had carried out by the lowest part of the mortal soul, Galen attributed to *Nature* (*physis*).[33] The spiritual or vital functions, including the emotions, both Plato and Galen attributed to the higher part of the mortal soul, in the heart. Although Galen attributes to Nature (*physis*) what Plato had attributed to one part of the mortal soul, Galen's divisions are the same as those of Plato, and his account thus assumes that the human body is the product of design in the same way that Plato does.

Galen himself saw no conflict between these two approaches, *both* of which were basic to his anatomizing. On the one hand there was a way of *learning* about the human body by dissection, which looked primarily at how the muscles were built on the frame of the skeleton, how the three systems of vessels were distributed throughout the body, and then moved on to inspect the viscera. This concerned *anatomy as structure*, and of course with its emphasis on the muscles and vessels was of especial concern to the practitioner as surgeon. And on the other hand there was *anatomy as key to function*, and here Galen, the supreme practical anatomist, willingly adopted the anatomical schema of Plato, the supreme theoretical – indeed fantastical – anatomist. Obviously Galen found this way of approaching the functioning of the inner organs of the body very valuable in his role as physician, and felt that it made coherent sense with his practical understanding of anatomy.

It is evident, therefore, that in his anatomy Galen is a Platonist in both important respects. That is, firstly his concern is with anatomizing the body of *man*, not with Aristotle's universal, 'The Animal'. Secondly, he sees man's body as physically demarcated into three areas, corresponding to the faculties of the soul. Hence for Galen, the diaphragm is an important boundary in the body. Moreover, this meant that Galen saw the heart and lung, both of them located in the thorax between the diaphragm and the neck, as both being *respiratory* organs, linked together by function. As we shall see later, this was to be of great significance. And in the third place, Galen sided with Plato (and against Aristotle) on issues such as the role of the brain as the source of the nerves, and on aspects of the doctrine of generation. In particular, although Galen certainly saw the heart as a very important organ, for him the brain was more important as the source of the nerves, and hence of movement.

The second major anatomical work by Galen, *On the use of parts*, though it follows the same sequence as *On anatomical procedures*, is of quite a different kind. Galen wrote it in the capacity of 'philosopher-physician'. It is, typically for Galen, devoted to showing that certain other people were wrong; in particular those who claimed that the world was simply a product of random chance, that is the atomists. Galen's other enemies on this occasion were those anatomists who did not pay sufficient attention to the 'uses' of the parts; and this meant primarily the long-dead Erasistratus and his live followers. For Galen, in this work, showing that each and every one of the parts of the human body has its 'use' is the key argument to knock down the claim that the world is the product of chance. Two enemies with one blow! The thesis that Galen maintains in *On the use of parts* is that all the parts of the human body were made for a purpose; that they were made in the best way to fulfil that purpose; that this purpose is the evidence for, and the expression of, Nature's foresight and wisdom; and that anatomizing to reveal the *uses* of the parts is the key to both understanding and appreciating that purpose and hence the wisdom of Nature. Here is Galen in full flow on this theme:

> when anyone looking . . . with an open mind sees that in such a slime of fleshes and juices [i.e. the body of man] there is yet an indwelling intelligence and sees too the structure of any animal whatsoever – for they all give evidence of a wise Creator – he will understand the excellence of the intelligence in the heavens. Then a work on the usefulness of the parts, which at first seemed to him a thing of scant importance, will be reckoned truly to be the source of a perfect theology, which is a thing far greater and far nobler than all of medicine. Hence such a work is serviceable not only for the physician, but much more so for the philosopher who is eager to gain an understanding of the whole of Nature. And I think that all men of whatever nation or degree who honor the gods should be initiated into this work, which is by no means like the mysteries of Eleusis and Samothrace. For feeble are the proofs that these give of what they strive to teach, but the proofs of Nature are plain to be seen in all animals.[34]

It might seem thus that we have a book of anatomical facts or information in *On anatomical procedures*, and a work in which certain anatomical facts are being put to a strange (and perhaps even unworthy) purpose in *On the use of parts*. This was not the view of later anatomists, who referred to both books as equal authorities on Galen's view of how the human body is and works. And it is certainly the case that the teleological view that Galen is arguing for in *On the use of parts* – the view that every part of the human body has a goal (*telos*) which it is constructed to fulfil – also underlies his whole enterprise in *On anatomical procedures* too. Indeed this is one of the ways in which Galen's whole approach to medicine is completely informed by the Greek rationalist tradition, especially from Aristotle: the belief that, by disciplined reasoning, we can discover function from structure and structure from function – because every structure has its particular form *in order to* fulfil the goal of its function.

Although, as mentioned above, the opinions of Herophilus and Erasistratus can be reconstructed only from references in the writings of Galen and others, and it is therefore difficult to be certain about what they thought,[35] nevertheless the hostile comments on Erasistratus by Galen certainly indicate that Galen saw a different body from the one that Erasistratus had seen. For although Herophilus and Erasistratus were already Ancients, having been dead for about five hundred years, their followers walked the streets of Rome of Galen's day and challenged his theories with their own. Galen's objections to Erasistratus's doctrines should probably be read therefore as being as much against Galen's own contemporaries as against the long dead Erasistratus. Galen's first ground of objection to Erasistratus was that he had held an atomistic view, and thus had neither sought nor believed in an Intelligence which designed the human body: he did not see the human body as the highest evidence of design. If this was the case, then nothing could have been more calculated to arouse Galen's anger. The second ground of objection that Galen seems to have had arose from his own ambition to create an account of the operations of the human body to be based on 'faculties' and humours, to replace those held by rival physicians of his day. In order to do this, Galen found it essential to show that certain claims put forward by the Erasistrateans were incorrect. Above all, in his desire to promote a form of therapy based on plethora and its cure by blood-letting, Galen required that the arteries be full of blood and not simply full of *pneuma* or life-giving air as Erasistratus had apparently maintained.[36] Galen's approach also required that the heart not be simply a muscle, as Erasistratus appears to have believed, but to consist of a special and unique substance with a 'pulsatile faculty'. Similarly, Galen needed it to be the case that the pulse of the arteries is transmitted from the heart through the coats of the arteries, whereas Erasistratus seems to have held that it was the impulsion of the pneuma from the heart into the arteries which caused them to pulsate.[37] With respect to the action of the heart, Galen also required the mitral 'valve' to be non-competent – where Erasistratus seems to have claimed it was competent – in order to allow a certain limited two-way flow in the blood vessel connecting the lung to the left side of the heart.[38] It was primarily to show such things that Galen used to perform animal vivisection in public, and why he claimed that his demonstrations compelled consent. For Galen this was the primary role of vivisection – as a means of refuting the conclusions of others. It was not his primary mode of anatomical investigation, however, as it seems have been for Herophilus and Erasistratus.

These then are the Ancients of anatomy, and all of them are in the 'rationalist' tradition. But while they were in the same tradition, and while Galen for instance looked back with respect to Plato, to Aristotle and (to a lesser extent) to Herophilus, there were major differences in the projects each of these Ancients had been engaged in. The most important difference of all was their objective in pursuing anatomy in the first place, and this allows us, in summary, to distinguish four different anatomical projects carried out by different Ancients:

1. Plato: concerned with the *human* body as the lodging of the immortal soul of man; sees man's body as deliberately shaped and arranged to fulfil the aims of the gods; man's body divided into three divisions, corresponding to the faculties of the soul; does not require engagement in dissection or vivisection.
2. Aristotle: the anatomizing of a very wide range of *animals* in order to understand the operations of the soul of 'The Animal'; an ongoing project, at least in Aristotle's hands. Man is just another animal, not the centre of the anatomist's attention. The heart is the central and most important organ in the body, the seat of the vegetative soul.
3. Herophilus and Erasistratus: *vivisecting* the *human* body in order to understand the operation of the human body for medicine.
4. Galen: like that of Herophilus and Erasistratus in that it is concerned with finding out about the operations of the *human* body for medicine, but engaged in by dissecting *animals* used as substitutes for his real object of study. Follows Plato's basic view of the functional divisions of the body; explicitly opposed to Erasistratus on a number of points. Has both a *structural* and a *functional* role for anatomy and anatomizing: the skeleton, the muscles, and the vessels, contrasted with the three divisions of the body and their contained viscera.

These are differences in *project*, not just in available creatures to dissect, and they give one different views of the body under scrutiny – such that these anatomists see *different bodies*. Hence Aristotle's view of the body was different from that of Plato, Herophilus's and Erasistratus's was different from that of Aristotle, Galen's different from that of Herophilus and Erasistratus. Most important of all, it was to turn out, was the difference between the views of Aristotle and Galen.

Galen's view of the body was based on Plato's view of the *human* soul, and this human body had *two* systems of blood vessels – the arteries based on the heart and the veins based on the liver, and also a system of nerves based on the brain. It is also important to note that Aristotle's view of the vegetative soul is quite different from the distinction later made by Galen (again based on Plato) between a nutritive and a vital soul in the trunk of the body, whose domains are divided by the diaphragm. For Aristotle, and also for his later followers, the vegetative soul reigns in the *whole* of the trunk of the body. Thus, for an Aristotelian anatomist, to study the heart is to study the operations of the vegetative soul, and similarly to study nutrition or generation is also to study the operations of the vegetative soul. Thus Aristotle's view of the (animal) body is not strictly comparable at all with Galen's view of the (human) body.

We shall later be concerned with when, why and how these different anatomical projects of the Ancients came to be recognized and brought back into practice. But for more than a thousand years, from the death of Galen in around 200 AD to about 1500, the differences between the anatomical projects of these Ancients was not seen; instead the Ancients, and especially Aristotle and Galen, were treated as having been engaged on one and the same anatomical project, one project with different aspects. Though Aristotle was sometimes referred to simply as The Philosopher, and Galen simply as The Physician (or the Prince of Physicians), both of them were regarded as equal authorities with respect to anatomy, and it was believed that what each of them had said about the (different) bodies they had seen was compatible or at least conciliable, and that their differences of emphasis or fact simply lay in their respective concerns as the Philosopher and the Physician.

Notes

1. This is the splendid translation by Benjamin Jowett in *The Dialogues of Plato, Translated into English* (first publ. 1871), Oxford Clarendon Press, 5 vols, third ed. 1892, vol. 3, pp. 491–92 (69b–71a). For a scholarly account of Plato's views on the soul see T.M. Robinson, *Plato's Psychology*, University of Toronto Press, 1970.
2. F.M. Cornford, *Before and After Socrates*, Cambridge University Press, 1979 (first published 1932), p. 59.
3. The standard Greek and English edition of Aristotle's works is the Loeb edition, published by Harvard U.P. and Heinemann of London. It is customary to refer to Aristotle's text by page and line following Bekker's Berlin edition of 1831, which are marked in all modern editions.
4. In taking *De Anima* as Aristotle's logical, if not chronological, starting-point for so much of his philosophy, I follow the opinion of a number of scholars, including the excellent and sympathetic account (dating from 1930) of Frederick J.E. Woodbridge, *Aristotle's Vision of Nature*, ed. with an Introduction by John Hermann Randall, Jr, Columbia University Press, 1965. For more recent work, see G.E.R. Lloyd and G.E.L. Owen, eds, *Aristotle on the Mind and the Senses, Proceedings of the Seventh Symposium Aristotelicum*, Cambridge University Press, 1978.

5. On the relation of Aristotle's philosophy to Plato's I follow Cornford, *Before and After Socrates*.
6. Cornford, op.cit., p. 64.
7. Marjorie Grene explains their relative positions thus: 'the core of Plato's doctrine, the Forms, recollection, the dualism of soul and body, Aristotle was at pains to refute . . . [Aristotle agreed with Plato that] it is forms that the knowing mind properly and rightly knows. Further, form is causal, it is the reason why things are what they are and the reason why our minds can know them. Further still, form is the source of unity: it is one as against the multiplicity of the informed, or of the unformed. Moreover, the mind in knowing form is somehow like it, at one with it. All this is common ground. But Aristotle found form, intelligibility, definiteness, where Plato had never found it: in the limited, recurrent but orderly processes of nature itself', *A Portrait of Aristotle*, London, Faber and Faber, 1963, p. 65.
8. Translations of Aristotle here are by Christine Salazar, based on those of the Loeb edition, and have been made specially.
9. 413 a 20.
10. 413 a 2; Woodbridge, op.cit., pp. 31–3.
11. Compare this with Plato, who believes he already knows the attributes of the soul, and then speaks of its/their anatomical *distribution* in the body of man.
12. 413 a 20–8.
13. 413 b 5.
14. 413 b, 414 a, 415 a.
15. 415 a 14–22. 'It is this cooperation of powers with a correlative field that Aristotle tries to work out in detail. This he does, first, by distinguishing powers such as appetite, reproduction [= generation], perception, imagination, memory, reason, and secondly, by analyzing the field in which they operate . . . Yet the soul is a unitary power. The exhibition of its activity as distinguishable powers is due to the fact of different fields of action and organs appropriate to them', Woodbridge, op.cit., pp. 46–47.
16. Historians have long believed that the animal books as we have them are in an unfinished or unpolished state, but if they are considered as aspects of the anatomical project discussed here, they have a greater coherence than hitherto suspected. See D.M. Balme, 'Aristotle's Use of Differentiae in Zoology', first published in 1962, reprinted in J. Barnes, M. Schofield and K. Sorabji (eds), *Articles on Aristotle 1. Science*, London, Duckworth, 1975, pp. 183–93.
17. Balme, op.cit., p. 192, citing *History of Animals* 491 a 9. With respect to the relation of *Parts of Animals* to the investigation of soul, see especially 641 a 16–25.
18. See Balme, op.cit., esp. p. 185.
19. 644 b 22 – 645 a 37, and 645 b 15–20.
20. See esp. 467 b and 469 a. On Aristotle's view of the heart see James Rochester Shaw, 'Models for Cardiac Structure and Function in Aristotle', *Journal of the History of Biology*, 1972, 5, 355–88.
21. That is, it is (uniquely) a Form without matter; see *De Anima* Book III, c. 3–5.
22. W.B. Yapp, *Vertebrates, Their Structure and Life*, New York, Oxford University Press, 1965, p. 130; see also p. 110.

23. Celsus, *De Medicina*, Loeb edition, Harvard University Press and Heinemann of London, 3 vols, 1960–61, The Proemium, vol. 1 pp. 14–15. This passage sometimes came to be merged with one from Pliny's *Natural History* – to the effect in the 16th century that the Egyptian kings themselves were said to have done the vivisecting! Pliny was discussing a particular medicinal property of the radish, discovered 'in Egypt by the kings dissecting the bodies of the dead to investigate diseases' (*in Aegypto regibus corpora mortuorum ad scrutandos morbos insecantibus*); Book 19. 86.
24. On these, see Gary Rubinstein, 'The Riddle of the Methodist Method: Understanding a Roman Medical Sect', Unpublished Ph.D. Thesis, Cambridge University, 1985.
25. On Galen's presentation of himself as the rightful heir of Hippocrates, see Wesley Smith, *The Hippocratic Tradition*, Cornell U.P., 1979. Galen also claimed to have written a (now lost) treatise *On the Anatomy of Hippocrates*.
26. The standard printed edition of Galen's writings in Greek and Latin is C.G. Kuhn (ed.) *Galeni Opera Omnia*, Leipzig, 1821–33, 20 vols. It was reproduced in microfiche by Oxford Microform Publications in 1976; this includes a useful guide (including to translations) by Vivian Nutton.
27. These have been translated as follows: *On Bones for Beginners* by Charles Singer, in *Proceedings of the Royal Society of Medicine*, 1952, **45**, pp. 767–76; *On the Anatomy of Muscles for Beginners*, by C.M. Goss, in *Anatomical Record*, 1963, **145**, pp. 477–501; *On the Anatomy of Veins and Arteries*, by C.M Goss, in *Anatomical Record*, 1961, **141**, pp. 344–366; and *On the Anatomy of the Uterus* by C.M. Goss, in *Anatomical Record*, 1962, **144**, pp. 77–83.
28. Book 1, c.2 (Kuhn vol 2, p. 220), translated specially by Christine Salazar; cf. note 34 below.
29. Ibid. (Kuhn vol 2, p. 226).
30. There is an English translation of the first nine books by Charles Singer, *Galen on Anatomical Procedures*, Oxford U.P. for the The Wellcome Historical Medical Museum, 1956. The remaining books have been translated from the Arabic by W.L.H. Duckworth as *Galen on Anatomical Procedures: The Later Books* (ed. by M.C. Lyons and B. Towers), Cambridge University Press, 1962; in these Galen continued on the brain and its dissection, on the sense organs of the head, on the larynx, on the generative organs, on the veins and arteries, on the cranial nerves and, finally, on the spinal nerves.
31. Galen, *On the Natural Faculties*, opening passage (Kuhn vol. 2, p. 1), translated specially by Christine Salazar; cf. the translation by A.J. Brock, London, Loeb Classical Library, 1916.
32. Galen acknowledges the *Timaeus* account as authoritative with respect to his own, he discussed it extensively and even intended to write a commentary devoted excusively to the medical sections of the *Timaeus*. See *Galen: On the Doctrines of Hippocrates and Plato*, ed., translated and commented on by Phillip de Lacy, Berlin, Akademie-Verlag, 2 vols, 1978–80 (*Corpus Medicorum Graecorum*), see esp. books 6, 8 and 9; this work has as a major theme the defence of the tripartite soul (Plato) against the single soul (Aristotle).
33. Galen may have taken the term here from the Stoics: see ibid., vol. 2, p. 375.

34. This is taken from the excellent English translation by Margaret Tallmadge May, *Galen On the Usefulness of the Parts of the Body*, Cornell University Press, 2 vols., 1968, vol 2, p. 731.
35. What was known about them was collected by J.F. Dobson in two articles: 'Herophilus of Alexandria', *Proceedings of the Royal Society of Medicine*, 1925, **18** (Section on History of Medicine), pp. 19–32, and 'Erasistratus' in the same journal, 1927, **20**, pp. 825–32. The 'fragments' of Herophilus have been assembled in Heinrich von Staden, *Herophilus: The Art of Medicine in Early Alexandria*, Cambridge University Press, 1989, but those of Erasistratus still await their compiler. On the dangers and problems of assembling such 'fragments' into coherent doctrines, see Harold Cherniss, *Aristotle's Criticism of Presocratic Philosophy*, Johns Hopkins Press, 1935, repr. 1983, Preface. Wesley D. Smith has some robust comments on the historiographical problems involved in such reconstruction in his 'Erasistratus's Dietetic Medicine', *Bulletin of the History of Medicine*, 1982, **56**, pp. 398–409. General writings on Herophilus and Erasistratus include Leonard G. Wilson, 'Erasistratus, Galen and the *pneuma*', *Bulletin of the History of Medicine*, 1959, **32**, pp. 293–314; James Longrigg, 'Superlative Achievement and Comparative Neglect: Alexandrian Medical Science and Modern Historical Research', *History of Science*, 1981, **19**, 155–200, and 'Anatomy in Alexandria in the Third Century B.C.', *British Journal for the History of Science*, 1988, **21**, pp. 455–88. The main works in which Galen attacked Erasistratus on respiration and the pulse, have been translated in David J. Furley and J.S. Wilkie, *Galen on Respiration and the Arteries*, Princeton University Press, 1984. Very valuable on all these matters is Jerome J. Bylebyl, 'Cardiovascular Physiology in the Sixteenth and Early Seventeenth Centuries', Unpublished Ph.D. Thesis, Yale University, 1969, Chapter 1.
36. On Galen's promotion of blood-letting, see Peter Brain, *Galen on Bloodletting*, Cambridge University Press, 1986, *passim*; on Galen attacking Erasistratus' views see also Charles Singer, *Galen On Anatomical Procedures*, Oxford University Press for The Wellcome Historical Medical Museum, 1956, pp. 176, 197–200.
37. For Galen showing vivisectionally that the heart and arteries pulsate simultaneously, see Singer, op.cit., pp. 194–96.
38. I put 'valve' into inverted commas here, since Galen regarded this structure as allowing two-way flow, whereas our modern concept of a valve is of something defined by the fact that it allows only one-way flow; i.e. Galen did not have, and was not using, our concept of 'valve'.

CHAPTER TWO

Between Ancients and Moderns

(i)

After the death of Galen it is not until the twelfth century that we find the practice of dissection again in the West. This practice of dissection was associated with the teaching of medicine at Salerno, in the period c.985 to c. 1225; the writings about anatomy which survive from Salerno date from 1100 to about 1225.[1] Here at Salerno, it appears, anatomies were conducted, but not on the human. Instead they were conducted on pigs:

> Because the structure of the internal parts of the human body was almost completely unknown, the ancient physicians, and especially Galen, undertook to display the positions of the internal organs by the dissection of brutes. Although some animals, such as monkeys, are found to resemble ourselves in external form, there are none so like us internally as the pig, and for this reason we are going to conduct an anatomy upon this animal.[2]

That is how one such Salernitan anatomy starts. It was a deliberate choice to dissect an animal rather than the human for, as another of these Salernitan anatomies says,

> Galen, a remarkable physician of those times, composed two books of anatomy of the dead and one of the living; but such treatment of the human body came to be considered inhuman, especially by Catholics, and the practice of dissection was transferred to animals.[3]

Hence the pig was chosen. So this dissecting was a practice, an avowedly Catholic practice, in which the human body was *avoided* as a subject for dissection, even though it was knowledge of human anatomy that was being sought. This view, that the dissection of the human body was abhorrent, was also one held by at least two noted Fathers of the Church, Tertullian and St Augustine.[4]

What is striking about these Salernitan anatomies is that the knowledge being sought in dissection was of immediate medical concern: the Salernitans concentrate very largely on which diseases particular parts are prone to suffer. Teleological discussion is almost totally lacking. The sequence of anatomizing the Salernitans adopted was: heart and related parts, then liver, then brain. And when they discuss anatomy without performing dissection, the order they prefer is, brain, heart, liver.[5]

This Salernitan form of anatomizing was replaced by one introduced in the medical faculty at Bologna, to which we shall turn in a moment.

(ii)

For about six hundred years in the universities of the West, from the mid 1200s until the early 1800s, anatomy was taken to be part of a larger study known as 'Natural Philosophy' (*philosophia naturalis*). Natural Philosophy was the branch of academic study concerned with Nature. We must not confuse the (now defunct) practice of Natural Philosophy with today's practice of science. All the people in this period whom we might want to classify as early 'scientists', would have actually called themselves 'natural philosophers' and their enterprise in discussing or investigating Nature they would have called Natural Philosophy; and they would have been right to do so. Natural Philosophy – all its departments – was about God. To study God through His works was the point of the whole enterprise.[6] Natural Philosophy studied the world – that is, nature – as created by God: it was nature studied *as creation*. Natural Philosophy was about God throughout its entire career, from its origin as a study in the mid–1200s right up to the nineteenth century when it was abandoned in favour of a new study, 'science'.[7] In being centrally about God and God's creation, Natural Philosophy as a study of Nature was quite unlike science today, even though science too is a study of Nature.

As a part of Natural Philosophy, anatomy too was therefore also about God. It was also of course relevant to the study of medicine, as conducted in and out of universities; indeed, over the centuries anatomy came more and more to be seen as important for the study of medicine. But even as studied for medicine, anatomy never ceased to be centrally about God. Anatomy was not a study which was primarily of value to medicine, and also about God. It was the other way about: it was primarily about God, and also of value to medicine. Thus at no time in the period from c. 1250 to the nineteenth century was anatomy a matter of so-many discrete 'facts' which had to be learnt or memorized. The teaching of anatomy certainly had the conveying of certain discrete information (as we might care to call it) as one of its goals, but the teaching of anatomy – even when conducted in a medical faculty – was centrally about God because it was about God's creation. The lessons to be learnt from anatomy were about more, much more, than the layout and working of the human body. This constitutes a radical difference in the goals of anatomy-as-Natural-Philosophy (as we shall be seeing it here), and anatomy-as-science (as we know it today), even if the same body, the human body, may be under observation in both practices. It also means that when actual anatomical investigations were pursued in the period of Natural Philosophy (i.e. *c.* 1250-*c.* 1800/1850), that 'research' in anatomy was primarily research about God's creation in order to discover or demonstrate things about God and His activity in this world. All the anatomy we shall be concerned with here was, in this sense, about God. Similarly all the anatomists we shall be dealing with were concerned with God in their anatomizing, in a sense far stronger than merely being religious men in a religious age.

Natural Philosophy was first created in the early 1200s in order to fight a particular heresy – a deviation from conventional Roman Catholic belief and practice. The heretics were the Cathars, who claimed that there were two Gods, an evil one and a good one. Matter was the creation of the evil God, and therefore was itself evil. The Catholic response to this had been to argue *from* created things (matter), showing that they were good, *to* the Creator (the single God), who must therefore also be good. Why this heresy was taken to be such a threat that it needed the creation of Natural Philosophy to stop it ever arising again, is discussed elsewhere.[8] For present purposes all we need to note is that the central goal of this new subject, Natural Philosophy, was to show the goodness of God's creation, and the constant presence and activity of this good God in His creation. The heretics had been arguing the opposite: that Nature was essentially corrupt and evil because it was not the creation of a good God, but of an evil principle.

The defenders of orthodox Catholic Christianity, especially the Orders of friars, successfully attempted in the early 1200s to infiltrate those new institutions, the universities or *studia*, and to influence their curricula in order to make them strongholds of orthodoxy. And to make the argument that God is good and that His creation is good in every particular, they turned to materials from ancient Greece, to works produced by Greek philosophers, and above all to the works of Aristotle: to his books on the heavens, on physics, on animals and so on. Such works were particularly attractive for this purpose because in them (as we have already seen) Aristotle had (a) discussed Nature (*physis*) as something separate from man, and capable of being investigated; (b) discussed it in order to show its *purposefulness*, to show the presence and operation of mind, intention and design in Nature; and (c) laid particular stress on the final causes of things – the 'final cause' of an object or a phenomenon being the answer to the question 'for the sake of what?' is a given thing or phenomenon in Nature the way it is. Works about Nature which put forward such an argument were exactly what the defenders of Catholic orthodoxy such as the Dominican friars Albertus Magnus and St Thomas Aquinas were looking for, and their suitability for this thirteenth-century task was one reason why the works of Aristotle were hunted out and translated from Greek or Arabic into the working language of the universities, Latin. To ensure that the message of these works came out loud and clear, the defenders of orthodoxy also began to offer extensive commentaries on the works of Aristotle. This is, at its very briefest, how Natural Philosophy was created and how the nature books of Aristotle became the mainstay of the Natural Philosophy curriculum of every university of Europe, and remained so until the eighteenth century.

In these respects anatomy was simply like any other Natural Philosophy study in this period. But there were of course more precise ways in which anatomy, as such, was about God and served the purposes of Catholic orthodoxy. We have already seen that Aristotle himself wrote extensively about animals and their anatomy: the works *History of Animals*, *Generation of Animals*, *Parts of Animals*, and others. We have also seen that Galen though not a Christian himself, wrote a great deal on anatomy in the second century of the Christian era. In doing so, Galen was happily following Aristotle in many respects, as he himself frequently says. In a crucial respect Galen followed Aristotle precisely: in his obsession with and pursuit of the 'final cause' of the (human) body and each of its organs and parts. The claim that there is design in Nature and that anatomizing shows that design, that structure is the key to function and function is the key to structure, is precisely the argument of his great work *On the use of the parts*, as we have already seen. Thus the animal books of Aristotle and the anatomical writings of Galen were the perfect sources for learning, teaching and pursuing anatomy in Europe of the thirteenth and later centuries where it was taken to be of central importance to claim that God was good and His creation was good. For such books could be used to show that the structure and operations of even the smallest and least significant feature of the anatomy of man and animals was the product of intelligent design and acted to bring about intelligent and good goals. To show this had, after all, been the intention of their ancient writers, even though they had not actually been Christians (either because they lived in the wrong place or at the wrong time or both). As Galen had said on this very issue:

> In fact you must not suppose that such skill as I have been explaining in this book [i.e., *On the use of parts*] is displayed in man alone; on the contrary, any other animal you may care to dissect will show you as well both the wisdom and skill of the Creator, and the smaller the animal the greater the wonder it will excite, just as when craftsmen carve something on small objects. There are such craftsmen even now, one of whom recently carved on a signet ring Phaethon drawn by four horses, each with its bit, mouth and

> front teeth so small that I did not see them all except by turning the marvel around under a bright light . . . Yet not one of these displays more perfect workmanship than the leg of a flea, and besides, the whole leg of the living, nourished, growing flea is pervaded by skill . . . If the Creator's skill is such when displayed incidentally, as one might say, in insignificant animals, how great must we consider his wisdom and power when displayed in animals of some importance![9]

The view of the body of man as a microcosm of the macrocosm – that is, a small version of the universe as a whole – was also important in the thirteenth century in approaches to its study. This ancient view came from the Platonic and neo-Platonic tradition, and now came to be interpreted in a specifically Christian way by reference to certain biblical passages and expressions. Man is a little world, a *minor mundus*; he is every creature, in the sense that he contains all things; man is created in God's own image; man is the last thing created, and hence is the highest likeness of God and the summation of all other creation, and he is the end or goal of the creation process; man stands between visible creation and the things which can only be grasped by the intellect – he is the mean. As a thirteenth-century English bishop expressed it in a sermon:

> Man is called 'the creation of the world', since in some way he is the end of all things . . . or is called 'the creation of the world' since in him in some way the whole world is placed, as the philosophers called him a smaller world; for this 'creation of the world', man, through the things which have been made – that is, through creatures – is able to see the invisible things of God.[10]

Man is and contains all things: to study him, his body and soul, is to study God's creation in summary, to study God's highest creation, to study the image of God.

That the anatomical works of Aristotle and Galen seemed tailor-made for the demands of the new Natural Philosophers of the thirteenth century is not ultimately too surprising, for the Christianity in whose defence they were now to be used was in large part originally produced by that same Greek culture.[11] And although Plato's influence, through the neo-Platonists, perhaps had more direct effect on the original shaping of the doctrines and claims of Christianity than those of Aristotle, yet here Plato too had preached the same message: the purposefulness of Nature, that Nature is the product of design. The essential goodness of the universe, the existence, adventures and afterlife of the soul, were characteristic of the beliefs of Plato and Aristotle and their ancient followers, such as Galen. Welded onto the teachings of Judaism, these concepts of the Greek philosophers came to be characteristic of that new hybrid religion, Christianity. Thus when our Christian anatomists much later discussed the operations of the body as being expressions of the soul (or of the souls, or of different aspects of the one soul), their vocabulary and concepts were common both to the Christianity in whose defence and for whose maintenance and strengthening their anatomizing was ultimately being undertaken, and to the Greek philosophers and physicians whose works they were using to undertake it. Quite rightly they saw little distinction between the Greek enterprise of 'philosophy' where it touched on Nature, and their own – Christian – enterprise of Natural Philosophy. And they were not confused about the difference between philosophy and religion: they simply drew the line in a different way than we do today.

(iii)

The first systematically natural philosophical form of anatomizing was introduced at Bologna, and probably before 1316, by Mundinus (Mondino de' Liuzzi) with his book *Anatomia*.[12] In this form of anatomizing, unlike that of Salerno, the human body is what is dissected. Mundinus was professor of practical medicine at Bologna (he was born *c.* 1270 and died in 1326). Mundinus's was to be the model for all anatomizing for two centuries, and his text was still the central text for those doing anatomical demonstrations on the human body well into the 1500s; it was this which was eventually to be challenged and replaced by alternative approaches in the Renaissance.

It is significant that this form of anatomical demonstration began at Bologna, in the early years of the medical faculty there. Bologna had long been a centre for the study of both civil and canon law, and by about 1260 both the teachers and students of law there had organized themselves into formal associations: a 'college' of doctors or teachers of law, and a 'university' of students of law. Together they constituted the *studium* of Bologna. The college of doctors of law began to take on the role of granting degrees to qualified students. By the end of the century the town authorities of Bologna, recognizing the wealth that such teaching brought to the town, had begun to offer salaries for the teaching of law (professorships). It was in the last decades of the thirteenth century that the teaching of medicine and philosophy at Bologna began to take a similar route.[13] For the town of Bologna began to provide paid posts for teachers of arts and medicine by 1305, and we know that in 1324 there were four such professors.[14] The students tried to have some control over the appointments to these positions, though they did not pay the professors directly. There were two professors of philosophy, one of whom taught Natural Philosophy using Aristotle's works; there was one professor of astrology, a subject considered most important then for medicine; and there was one professor of practical medicine. It was this last post which Mundinus held in 1324, and had already held for some years.

Thus medicine on the one hand and philosophy and arts on the other came in together at Bologna. By 1268 the teachers of arts and medicine had formed a (single) 'college' and were examining students in medicine, and by about 1295 students of arts and medicine from outside Bologna had formed a (single) 'university of arts and medicine'. What this meant was that most of these students were students both of philosophy – and that now included Natural Philosophy – and of medicine at the same time and in the same faculty and studying for a degree in both subjects combined. This state of affairs was to continue for centuries at Bologna. The link between the discipline of Natural Philosophy and the discipline of medicine could hardly have been closer. And that meant that students of medicine would thus already be familiar at least to some extent, with Aristotle's book on the soul and his book (or books) on animals when they came to study anatomy in the context of medicine. They would read, interpret and understand that anatomy 'Natural-Philosophically', if one may coin a term.[15]

It was in this context that the *demonstration* of anatomy on the human body – an event thus both medical and philosophical – was introduced for students of medicine. Mundinus, as professor of practical medicine trying to teach the medical doctrines of Galen, regarded anatomical knowledge as fundamental. But Mundinus was teaching this Galenic medicine within the world of a *studium*, a world unknown to Galen or Aristotle. Hence he became aware that there was no appropriate text by Galen or Aristotle or an Arab physician for the form of anatomical demonstration he had in mind. So he created a demonstration-dissection exercise for which he had no ancient model.

The exercise of anatomical demonstration, as initiated by Mundinus, consisted of taking a human corpse and dissecting it in front of an audience, over three or more days. The body was that of a criminal who had just been put to death. Thus the body of the lowest form of human, an executed criminal, was used to show the highest handiwork of God. From his Avicenna and the writings of Galen which were available to him, Mundinus constructed a text which would be suitable for a three-day dissection sequence and which would deal with all the important inner parts as they came to light during dissection and mention certain features of them. What happened was that a professor (Mundinus or one of his successors) sat in his high professorial chair, or *cathedra*, reading out the Latin text of Mundinus slowly (and, if he wished, commenting on it). A surgeon performed the dissection, and someone

acted as an 'ostensor', pointing out the parts as they came to view. Although pictures which represent this event show the lecture and the demonstration taking place in the same visual space, it is quite possible that the two events were separated in time (see Figure 2.1).[16]
Later statutes at Padua University specify that:

> The rector and the *consilarii* are to depute one of the *doctores extraordinarii* who will recite and read the text of the *Anatomy* of Mondino, and one of the *doctores ordinarii*, either of practice, or of theory, who will explain the aforesaid text line by line, and what he has explained according to text and letter, let him demonstrate by visual testimony, and verify in the cadaver itself.[17]

Mundinus's work has traditionally been seen by historians as a good idea gone wrong. It was a good idea to actually dissect the human body for the benefit of medical students, to show them the inside of the body with which their practice would later deal. But it was a bad idea in that Mundinus's text was a terrible old mixture of false information from corrupt Galenic texts, and from the poorly translated texts of Arab physicians who, as Muslims, had themselves been prohibited from dissecting the human body. Worst of all was the execution of Mundinus's scheme: the professor sat in his high chair reading out the text, while a surgeon down below performed the dissection and an ostensor pointed out the parts as they were mentioned. This second person, the ostensor, was charged with finding in the body what was mentioned in Mundinus's text: 'and what he has explained according to text and letter, let him demonstrate by visual testimony, and verify in the cadaver itself'. What chance here of noting discrepancies between the text and the body, with the professor so far away and the operators so ignorant? What chance here, in a mere three-day dissection, and that in public, of discovering new and true things about the body, even those right in front of one's eyes? Constrained by the limited number of bodies available for dissection (one or two a year); constrained too, by the advancing putrefaction of the corpse, to dissect it in the order of corruption (belly, chest, head), Mundinus's anatomies had – historians have held – a stultifying effect for two centuries. They held up progress.[18]

Yet little attempt has been made by historians to see Mundinus's work as innovatory in its own time, or to try and understand why for two hundred years other people found Mundinus's text a good model for their own anatomical demonstrations. Obviously, if we see Mundinus's anatomizing as a modern teaching exercise with the aim of conveying certain discrete items of information, then it must have been of limited effect. If we see it as a 'research' exercise, in which discrepancies and novelties were being sought, then it must have been a failure. But it was neither. It was not intended to find fault with the ancient authors at all.

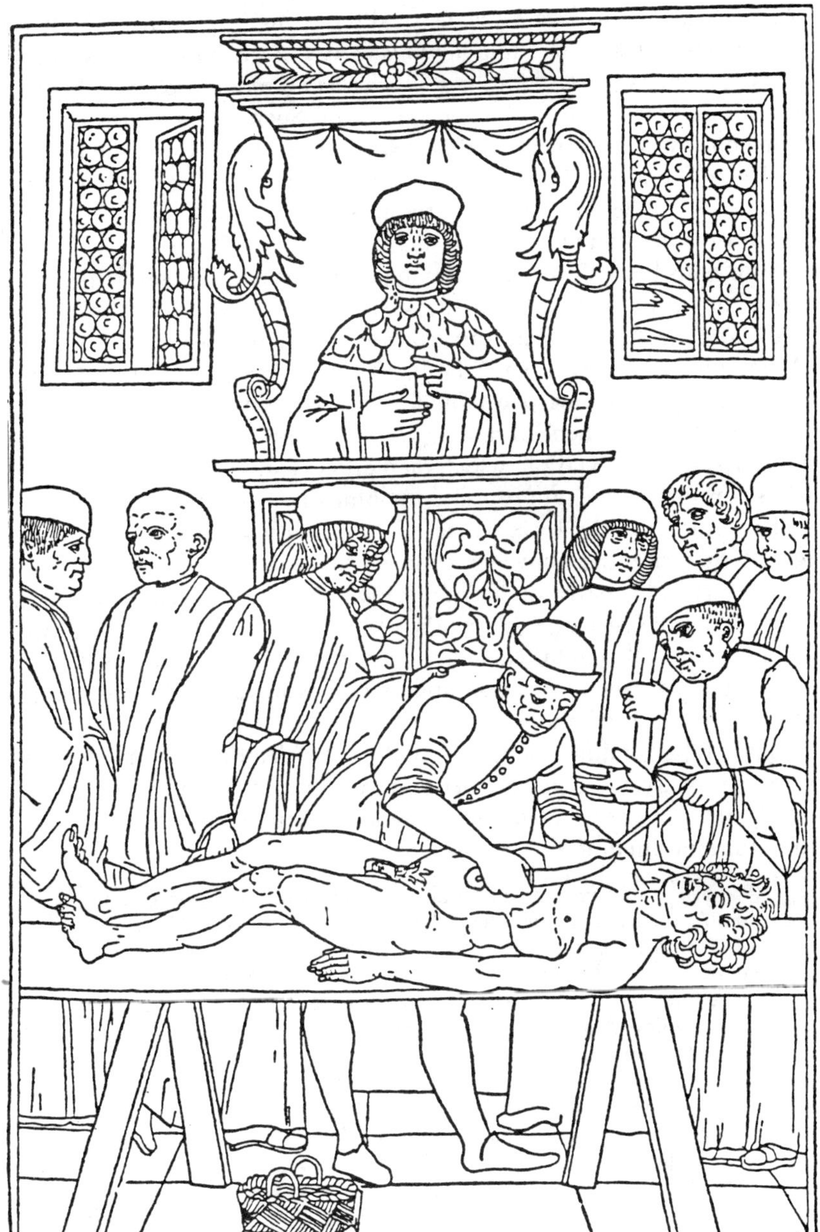

2.1 Anatomical demonstration scene, showing the division of roles between the professor (physician), the dissector (surgeon) and the demonstrator. From the *Fasciculo di Medicina*, Venice, 1493.

Nor was it intended as a means of discovering new things about the human body. It was instead a *demonstration* of the human body and its parts, a demonstration of a kind which has no modern parallel because it is an enterprise we no longer engage in. Such a demonstration had twin goals: it was medical, in that it was for the attention of the medical students; but it was also natural philosophical, in that it used the dissection of the human body to point to and admire God as the Creator of that body.[19]

Mundinus begins his text by saying that it has been composed for his pupils in medicine and is on the knowledge (*cognitio*) of the human body and its parts, which arises from anatomy. We need to know (Mundinus says) about man's body as a whole before we go on to the parts, for all our knowledge should begin from what is 'more known' to us. What this treatment of man-as-a-whole consists of for Mundinus is a discussion of the ways in which man differs from other animals. There is his upright stance, appropriate for him since (among other reasons) he has a most perfect form which he shares with the Angels and Intelligences that rule the Universe; thus all his senses are properly in the upper part of his body, on account of which man is called 'microcosm' or smaller world because, like the world, man has an upper and a lower. Man is unique also in his capacity to learn skills, in his lack of defensive parts (such as a shell or claws), and in his possession of that instrument of instruments, the hand.

How did Mundinus actually conduct his demonstrations? For the sake of brevity we may use here an engaging description, written in 1363, of how one of Mundinus's pupils conducted demonstrations in the Mundinus manner:

> We prove anatomy in bodies newly dead by decapitation or hanging. At the least [we deal with the anatomy of] the official members of the inner things; of the muscles of the flesh, of the skin, and of many veins and nerves, namely as to their origin, after the manner of Mundinus of Bologna treats, who wrote upon this and did it many times. And my master Bertruccius by this manner: a dead body, laid on a bench, he made of it 4 lessons.
>
> In the first lesson the nutritive members were treated, for they be more soon putrible.
> In the second, spiritual members.
> In the third, animate members.
> In the fourth are treated the extremities.
>
> And after [the manner of] the Commentator on *On the Sects*, in every member were 9 things to be seen: position, substance, complexion, quantity, number, figure, the shape and the fastening together, act and utility, and which be the sickness which may happen in it.[20]

Thus of every part seen, nine things are posited, which together serve to make up its anatomical identity. They include the action and use of each part. These nine are the 'observables' which, as a set of things to observe in anatomy seem to have been introduced by Mundinus.[21] So there was much information related at an anatomy, and in a consistent manner, of interest to the future physician.

The 'official members of the inner things' are parts which carry out some specific office or role for the good of the body as a whole. They are also called 'organs', *organa* in Greek meaning 'instruments'. The 'official members of the inner things' are, according to Mundinus, those which are ordained directly for the conservation of the individual. This conservation falls into three departments: the natural/nutritive, the vital/spiritual and the animal/animating, which correspond to three areas of the body – the belly, chest and head. Like a good follower of Galen, himself a follower of Plato in this respect, Mundinus sees these three regions as separated by two natural boundaries in the body. One of these boundaries is the diaphragm, for 'one of the uses of the diaphragm is that it should divide the spiritual members from the natural members'. The other natural boundary, dividing the spiritual members from the animal members, was for Mundinus (again following Galen and Plato) the neck; more precisely, it was the mouth and its contents.

Mundinus's view of the body as consisting of three 'bellies' or 'venters', each with a specific role in the life of the human, actually came from the writings of Galen himself – from *De iuvamentis membrorum*, the medieval heir to *On the use of parts* (*De usu partium*).[22] It would also have been available to him in a translation of Galen's *On the natural faculties* made from Arabic to Latin in the previous century, and known to Mundinus under the title *De virtutibus naturalibus* ('On the natural powers').[23] In turn, of course, it had come from Plato, as we have seen. The three-venter concept, however, had not been used by Galen as his basic procedure for dissecting – for *anatomy as structure* – but rather for *anatomy as function* (as discussed above in Chapter 1). So in adopting this for dissections, Mundinus was adopting a dissection sequence that Galen himself had not used.

Because Mundinus's account of the workings of the human body set the agenda for subsequent anatomical conceptualizing for over two hundred years, it is important for us to get some grasp of it here. In Mundinus's formulation (see Figure 2.2), built of course on Galen, the role of the lower venter is to *nourish* the entire body. The stomach receives the food and digests it into chyle. This is passed on to the liver via the portal ('gateway') vein, where the chyle is transformed into blood. This blood issues from the liver through the *veins* to every part of the body: this blood is what nourishes the body. Meanwhile the

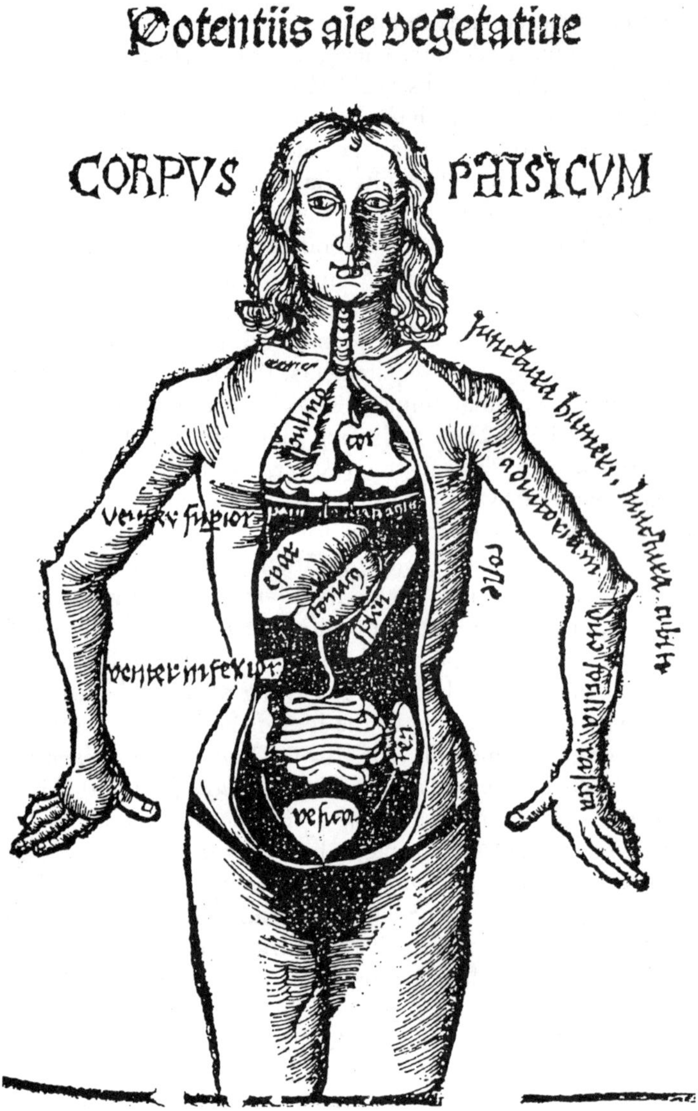

2.2 The diaphragm as a boundary between the lower and middle venters of the human body. The diaphragm is shown running horizontally across the body, just below the armpits. Below the diaphragm lies the lower venter ('venter inferior') and its contents (liver, stomach, spleen, kidneys, bladder). Above the diaphragm lies the middle venter ('venter superior') and its contents (heart and lung), and it is portrayed as continuing up to the top of the throat. From the section 'On the powers of the vegetative soul' ('De potentiis animae vegetativae'), in G. Reisch, *Margarita Philosophica nova*, Strasburg, 1512.

intestines and kidneys have their task of disposing of the wastes of these processes; and the spleen and gallbladder have their particular tasks of collecting the melancholic humour unsuited for evacuation, and cleansing the liver of bile, respectively. The food thus goes through two stages of elaboration (in the stomach, and then in the liver) before it becomes blood; there is a residue from this process which is discharged.

What happens in the *middle* venter is that the organs there 'vitalise' the body: they keep it alive in a way which food alone cannot do. The air passes in at the mouth or nose, down the windpipe (trachea) and hence to the lung. It is then drawn through the 'artery-which-looks-like-a-vein' to the left ventricle of the heart. Meanwhile the *right* ventricle draws 'natural' or 'nutritive' blood from the liver via the vena cava and (in some way unspecified by Mundinus) makes it somewhat 'more subtle'. On the contraction of the right ventricle, most of the blood then passes back through the way it had come, now subtilized for the nourishment of the organs of the body; it continues on its way to the body through the *veins*. A proportion of it is carried through the 'vein-which-looks-like-an-artery' to nourish the lung. And a further proportion of it is passed to the *middle* ventricle of the heart. Mundinus describes this 'middle' ventricle as 'not one but many small cavities', which get increasingly more fine; and by virtue of passing along these constantly narrowing little channels the blood receives a new preparation: it gets further subtilized, ready for it to be turned into spirit (see Figure 2.3). From the middle ventricle this small portion of blood now reaches the *left* ventricle. Here 'spirit is formed' from this blood 'by a strong subtilising and evaporating heat': in this process the air from the lung plays a highly important (but here unspecified) role. Mundinus probably understood it to be fuelling the heat, as in a furnace; other anatomists were to be more explicit about the role of the air in this process. Then, when the arteries expand and the left ventricle contracts, two things happen. The *spirit* formed from the blood moves into the *arteries* and thence to all the organs of the body; it is consumed in keeping the body alive. And the smoky vapours which are the residue of the spirit-generating process, are passed back to the lung and are then breathed out. Then the cycle starts again. As with the nutritive process in the lowest venter, there are two stages of preparation here (in the 'middle' and left ventricles), and there is a residue from this process which is discharged.

Finally, in the *highest* venter are the organs which are concerned with what we can here call 'animation'. Vital spirit from the left side of the heart moves up the arteries of the neck and reaches a network of small arteries at the base of the brain. This network of vessels, the *rete mirabile* or 'amazing net', makes the first elaboration by passing the

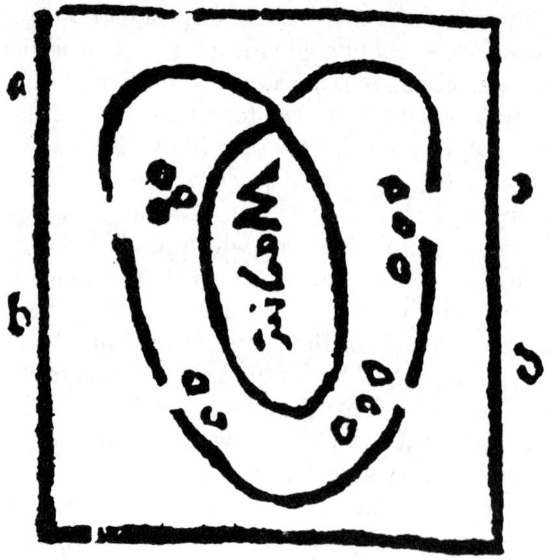

2.3 The three-chambered heart, as portrayed in Johannes Adelphus, *Mundini de Omnibus Humani Corporis Interioribus Membris Anathomia*, Strasburg, 1513.

vital spirit through successively narrower channels. Then the vital spirit passes to the large spaces within the brain – the ventricles (little venters or bellies) – where it receives its final elaboration. It is now blood in its most highly refined form: it is 'animal' spirit. This 'animal' spirit is what is poured down the nerves (from the rear of the double front ventricle) so that *sensation* can come via it from the sense organs to the brain. The animal spirits are also the agency of the nerves of *motion*, operated from the backmost ventricle of the brain. There are thus two stages of elaboration in the brain (in the *rete mirabile* and in the ventricles); again there is a residue which is discharged (down the nose). The substance of the brain is not active in this generation of the animal spirits, nor in the intellectual processes; it merely 'tempers' the vital spirit. Again we can see that everything in this venter, including the eyes and ears, contributes toward one kind of activity: the *animating* of the human body.

Each venter and its contained organs thus has a distinctive and unique role in this account of human anatomy. Yet of course connections between the different venters were recognized by Mundinus. For instance, the blood produced by the liver in the lower venter was understood to pass up the large vein (the vena cava), which penetrated the diaphragm, into the middle venter and come to the right side of the heart. Similarly vital spirit prepared in the (left) heart was understood to pass through the arteries up out of the middle venter to the highest venter, and thus to the *rete mirabile*. And animal spirit, in turn, was distributed from the brain to all the sense organs and muscles of the body through the nerves. However, although these and other connections were recognized, each of the major roles or functions (the natural, vital and animal) was deemed to take place *exclusively* in one of the venters. All the organs of nourishment are in the lower venter; all the organs of animation are in the highest venter; and all the organs of vitality are in the middle venter.

Mundinus's adoption of the three-venter division for both the dissection and the understanding of the human body carried with it certain consequences for how the inter-relation of organs could and had to be seen henceforward by anatomists. The very format of the anatomizing process meant that certain doctrines of Galen became inviolable and unquestionable. With respect to the heart, this meant that it had to be seen as an organ in *two* systems; and the left heart had to be seen as indissolubly associated with the *lung* and with respiration. Its primary importance in the eyes of Mundinus and other anatomists did not have to do with the blood, but with the air. The heart was the central organ of respiration, it was 'for the sake of' respiration; it performed perfectly the highly important role of vitalizing the body. And that was why it was in that very important place, the centre of the central venter. It was because they were following the route of the *air* that Mundinus and other anatomists always looked at the heart, the lung and the trachea (wind-pipe) together. Now, trachea is short for 'trachea arteria', which means 'rough artery'. We would today expect that to qualify as an artery, a vessel should contain blood. But the trachea had in fact been quite correctly originally defined in Antiquity as an artery, because it contained air (*aer*, which is the linguistic root of *arteria*). Similarly all the other arteries of the body had been first distinguished from the veins on the grounds that they contained air (in a certain form) because they were respiratory vessels, the vessels which transmitted vitality, life, to the body. This helps explain the curious names that the two vessels connecting the heart and the lung had acquired by Mundinus's day, the 'vein-which-looks-like-an-artery' and the 'artery-which-looks-like-a-vein'. For the first was, for Mundinus and his followers, a vein by function (for it distributes *nutritive blood* to the lung) but which has the behaviour and physical characteristics of an artery (pulsation, and with a thick wall); and the second was an artery by function (for it carries *air*) but which has the behaviour and physical characteristics of a vein (a thinner wall and no pulsation). For at root, the distinction between veins and arteries was always one of function: arteries were associated with respiration, veins with nutrition. And that is why the (left) heart, the lung, and the trachea were looked at as a *set* of organs, for they participated in one and the same function: respiration.

One can now see in all this where – and where alone – Mundinus and subsequent anatomists could draw, as it were, imaginary dotted lines around the organs of the body such that they demarcated things which (for them) had demarcation by Nature. Thus in the middle venter the heart-lung-trachea were, for them, unified by function, unified with the arteries. It was simply not possible for them to conceptualize the functioning of the (left) heart without the lung and trachea. It would require a change in anatomical practice – new 'spectacles' – before such doctrines could be questioned.

With Mundinus therefore what we find is that instead of shunning the human body as the object of dissection, as the Salernitans had done, the human body is precisely what he chose to dissect. The sequence of dissecting that Mundinus chose was not something simply dictated to him by the order of corruption, but something chosen through a desire to rise through the hierarchy of soul – from nutritive to spiritual to animal, from liver to heart to brain – that soul whose instrument, the human body, was being demonstrated. It is inherent in the deliberate adoption of the Plato-Galen division of the human body, that one is talking about the soul and about the hierarchy of soul. The experience of Salerno shows that it was possible to dissect for medical knowledge without invoking God or the soul, and that this involved a sequence different from that later adopted by Mundinus. But now with Mundinus the divine was primary and fundamental to anatomical demonstration and gave it its central identity. As Guy de Chauliac put it in 1363 in his 'Sermon universal of anatomy' as it was contemporaneously translated into English:

> For after Galen, the lantern of leeches [= the light of physicians], in the 17th book of the profit of particles [= *On the use of parts*], the last chapter save one: there are four profits or conditions of Anatomy.
>
> The first forsooth and most, to the might of God to be marvelled.
> The second, to be knowledgeable of the patient's parts.
> The third, to prognosticate of the dispositions to come of the body.
> But the fourth is to the curation of sickness.[24]

In the Mundinus anatomizing paradigm the first and greatest role of anatomizing is marvelling at the might of God.

The change we see here from Salernitan to Mundinian modes of anatomizing – encapsulated by the reintroduction of God and the soul –

was of a kind that could not have been due simply to the acquisition of a better range of ancient texts. Indeed Mundinus had access to only marginally better ancient texts. But Mundinus's introduction of the anatomical demonstration, and of this particular three-venter approach to performing it, was part of the introduction into the *studia* of thirteenth-and fourteenth-century Europe of Natural Philosophy, itself in part a response to the new pressures being experienced by Catholicism in Mundinus's own time. That dissection of the human body, hitherto so abhorrent to Catholics, was now being conducted freely in a Catholic world and under Catholic aegis, does indeed indicate that it was now performing a service for, or at the very least in accord with, Catholicism. It was anatomy as Natural Philosophy. Knowledge of God, the single, good, Catholic creator God, had been introduced as its primary goal: 'to the might of God to be marvelled'.

What then can we conclude about the body that Mundinus and subsequent anatomists saw, demonstrated and (on occasion) explored in their university context? That it was the human body, and that it was looked at primarily for the sake of illustrating the power and goodness of God.[25] Every part of that body, from the smallest to the largest, had its role or function, which the anatomist could become acquainted with. This body was divided into three venters, each of which had a general function presided over by a separate 'soul'; the roots of this lay in Plato's highly teleological and soul-obsessed story in the *Timaeus*. And the human body as a whole existed 'for the sake of' the immortal soul. Everything had its role and was in the best possible place. Everything worked together for good. The human body was being looked at as the microcosm, and as having been created by God as the high point of His creative act. It was being looked at as the temporary vehicle of the immortal soul. It was being looked at as the highest evidence of the existence of intelligent design in Nature – a particularly good argument against the anti-Catholic heretics of the medieval period.

Finally we need to note that, over and beyond Mundinus's use of anatomy as Natural Philosophical demonstration, there were certain major differences between Mundinus's project as anatomist and that of Galen. In the first place, although Galen did indeed talk extensively about the three-venter distinction, it had not been the basis of his own dissection practice when teaching others. For Galen started with the skeleton and bones, moved on to the muscles and then to the systems of arteries, veins and nerves (as we saw in Chapter 1). Thus Mundinus's approach inadvertently inverted Galen's preferred sequence of dissection practice; instead of demonstrating it first, Mundinus demonstrated the skeleton last – if there was time. There was a second difference between Mundinus's and Galen's anatomical projects: that Mundinus is dissecting the body of *man*, which Galen had never done. But because Mundinus was intent on following Galenic *doctrine* as closely as possible, this difference in anatomical subject made no significant difference to what Mundinus could see as being anatomically true.

In the light of our earlier discussion of the nature of Natural Philosophy, it ought now to be possible to appreciate that the practice of anatomical dissection as such was also a quasi-religious event in that it was a ritual or ceremonial activity with a message that pointed to God and His goodness. The conduct of a 'public' anatomy was highly formalized and ritualized, right through from the setting-up of the environment where it was to take place, to the re-assembly and burial of the body which had been dissected. Graphic portrayals of this event regularly presented it like a religious ritual. But, more than this, all thinking, talking or teaching about anatomy, to or by anyone whatever, had the persistent function of being about God and performed to His glory. In this sense all anatomizing in our period was as ceremonial and religious an experience as a church service: it was the unveiling of holy mysteries. Without such natural philosophical motivations to undertake it, without the functions it was believed to be performing for the defence of Christianity and Christian society, it may well be that anatomy would not have been undertaken by anyone at all in the West in our period. Hence when we do find anatomy being carried out in this period, we should expect it to be of a form shaped to serve these purposes.

Notes

1. On medicine at Salerno see P.O. Kristeller, 'The School of Salerno', *Bulletin of the History of Medicine*, 1945, **17**, pp. 138–94; Brian Lawn, *The Salernitan Questions: An Introduction to the History of Medieval and Renaissance Problem Literature*, Oxford, Clarendon, 1963. On anatomy at Salerno, see George W. Corner (ed. and translator), *Anatomical Texts of the Earlier Middle Ages, A Study in the Transmission of Culture*, Washington, Carnegie Institution, 1927 (reprinted New York, AMS Press, 1977), p. 51. All the Salernitan texts I am discussing are in this work.
2. Corner, op.cit., p. 51, as translated by Corner.
3. Corner, op.cit., p. 67. The Latin reads: 'Unde Galienus qui tunc temporis mirabilis existebat medicus duos libros de anathomia mortuorum, unum vero de anathomia vivorum somposuit [*sic*]. Sed quia inhumanum videbatur et maxime apud catholicos, corpus humanum tali modo tractari, usum anathomiae ad corpora brutorum transtulerunt'; from Franz Redeker, *Die 'Anatomia magistri Nicolai phisici' und ihr Verhaltnis zur Anatomia Chophonis und Richardi*, Inaugural-Dissertation, Leipzig, 1917, p. 30.
4. Tertullian, *On the Soul*, 10,4 and 25,5; Augustine *Concerning the City of God, Against the Pagans* Book 22, c.24 (all three passages are against Herophilus and vivisection).

5. Corner, op.cit., *passim.*
6. See Roger French and Andrew Cunningham, *Before Science: The Invention of the Friars' Natural Philosophy*, Aldershot, Scolar Press, 1996. Natural Philosophy should not be confused with the much later study, natural theology.
7. See Andrew Cunningham, 'Getting the Game Right: Some Plain Words on the Identity and Invention of Science' *Studies in History and Philosophy of Science*, 1988, **19**, pp. 365–89; Andrew Cunningham and Perry Williams, 'De-Centring the "Big Picture": *The Origins of Modern Science* and The Modern Origins of Science', *British Journal of the History of Science*, 1993, **26**, pp. 407–32.
8. See French and Cunningham, op.cit.
9. As translated by Margaret Tallmadge May, *Galen on the Usefulness of the Parts of the Body*, Cornell University Press, 2 vols, 1968, vol. 2, pp. 731–32.
10. This paragraph is indebted to James McEvoy, *The Philosophy of Robert Grosseteste*, Oxford, Clarendon Press, 1982, chapter 6 'The Place of Man in the Cosmos'; the quotation is from pp. 393–94, translated by me. In speaking of seeing the invisible things of God through the visible creation, Grosseteste is alluding to *Romans* 1.20. On how this is a specifically Franciscan approach see French and Cunningham, op.cit. Rudolf Allers, 'Microcosmus from Anaximandros to Paracelsus', *Traditio* 1944, **2**, pp. 319–407, is not of any help here. For the persistence of the microcosmos image, see Leonard Barkan, *Nature's Work of Art: The Human Body as Image of the World*, New Haven, Yale University Press, 1975.
11. See French and Cunningham, op.cit., Chapter 1.
12. Mundinus is discussed and his text translated in Charles Singer (ed.), *J. Ketham, The Fasciculus di Medicina, Venice, 1493*, Florence, Rihier, 1925, pp. 45–112. For the Latin text I have used Berengario da Carpi, *Carpi Commentaria cum Amplissimis Additionibus Super Anatomiam Mundini cum Textu Eiusdem in Pristinum et Verum Nitorem Redacto*, Bologna, de Benedictis, 1521.
13. For information on Bologna I have used Nancy G. Siraisi, *Taddeo Alderotti and his Pupils, Two Generations of Medical Learning*, Princeton University Press, 1981, Chapter 1.
14. Ibid., p. 21.
15. On the alliance of medicine and Natural Philosophy in such universities, see Jerome J. Bylebyl, 'Medicine, Philosophy and Humanism in Renaissance Italy', in John W. Shirley and F. David Hoeniger (eds), *Science and the Arts in the Renaissance*, Washington, Folger Books, 1985, pp. 27–49, esp. 27–34; and P.O. Kristeller, 'Philosophy and Medicine in Italy', in Stuart F. Spicker (ed.), *Organicism, Medicine and Metaphysics: Essays in Honour of Hans Jonas on his 75th Birthday, May 10, 1978*, Dordrecht, Reidel, 1978, pp. 29–40. For the curriculum at Bologna in 1405, where the relation of philosophy to medicine is evident, see Lynn Thorndike, *University Records and Life in the Middle Ages*, New York, Columbia University Press, 1944, pp. 279–81.
16. On interpreting such images, see Jerome J. Bylebyl, 'Interpreting the *Fasciculo* Anatomy Scene', *Journal of the History of Medicine*, 1990, **45**, pp. 285–316.
17. Statute of 1465 or before; see ibid., p. 310, and p. 309, n. 66.

18. The classic example of a historian who made such judgements is the late Charles Singer, himself the translator of Mundinus; see for instance his *The Evolution of Anatomy*, London, Kegan Paul, 1925. Such judgements are still often made today.
19. For a comparison with anatomy performed directly for the teaching of surgery, see the *Chirurgia* of Guglielmo da Saliceto, who 'unlike Mondino dei Liuzzi in the preface to his *Anatomia* written a generation later, made no attempt to present the study of human anatomy as inherently valuable because it gives knowledge about a noble part of God's creation': Nancy G. Siraisi, 'How to write a Latin book on surgery: organizing principles and authorial devices in Guglielmo da Saliceto and Dino del Garbo', in Luis García-Ballester, Roger French, Jon Arrizabalaga and Andrew Cunningham, eds, *Practical Medicine from Salerno to the Black Death*, Cambridge University Press, 1994, pp. 88–109; see p. 99.
20. Bjorn Wallner (ed.) *The Middle English Translation of Guy de Chauliac's Anatomy*, Lund, CWK Gleerup, 1964, pp. 58–9; I have modernized this passage. The Commentator on the Sects is John of Alexandria; for the commentary see C.D. Pritchet, *Johannis Alexandrini Commentaria in Librum De Sectis Galeni*, Leiden, Brill, 1982.
21. See R.K. French, 'A Note on the Anatomical *Accessus* of the Middle Ages', *Medical History*, 1979, 23, pp. 461–68. The 'observables' in anatomy derive from John of Alexandria's commentary on Galen's *De Sectis*, and ultimately from the categories of Aristotle. According to French, Mundinus 'took the *accessus* directly from John of Alexandria, and followed the *occasiones* for every organ in the body', ibid., p. 465.
22. For this see R.K. French, '*De Juvamentis Membrorum* and the Reception of Galenic Physiological Anatomy', *Isis*, 1979, 70, pp. 96–109. On the relation of Mundinus's views to those of Galen, see Jerome J. Bylebyl, 'Cardiovascular Physiology in the Sixteenth and Early Seventeenth Centuries', Unpublished Ph.D. Thesis, Yale University, 1969, Chapter 2.
23. On the availability of this, see French, op.cit.
24. Wallner, op.cit., pp. 54–55; I have cannibalized both versions here.
25. For an extended instance of this, see Manfredi's dedication of his *Anathomia*, in Charles Singer, 'A Study in Early Renaissance Anatomy, with a New Text: The *Anathomia* of Hieronymo Manfredi (1490)', in Charles Singer (ed.), *Studies in the History and Method of Science*, Oxford, Clarendon, 1917, pp. 79–164. Significantly Singer published the Latin text of this dedication but had almost none of it translated.

CHAPTER THREE

The Renaissance and Anatomy: The First Changes

(i)

We come now to the Renaissance and the fate of anatomizing in it. In and around the year 1500 the whole learned world in the West was in the throes of what felt at the time like the most exciting and far-reaching change for over a thousand years, whose point was to take a giant step backwards for mankind, to return to the Ancients of classical Greece and Rome. For half a century or more before 1500, in the economically dominant area of western Christendom, certain people in the rich and ostentatious city-states of northern Italy, and especially in Florence, Rome and Venice, had been engaged in a campaign. In sculpture, architecture, painting; in genres of writing; in language (classical Latin, and Greek); in mathematics and astronomy; in philosophy; in politics; in fact in every field of Ancient endeavour, they were campaigning to recover the principles and practices of the Ancients and restore them to use. In sculpture one could take as models surviving statues from Antiquity, as Donatello did; in architecture one could measure ancient buildings to recover the principles on which they had been built, as Brunelleschi did, or one could read and even rewrite Vitruvius's treatise *De Architectura* from the 1st century BC, as Alberti did, and seek to plan and build according to his guidance. In each case this activity restored a category of endeavour (sculptor, architect) and a theory-based practice (sculpture, architecture) which had been in abeyance for a millenium; and in each case this activity of deliberately reviving ancient practices in a modern world led, unintentionally but inexorably, not so much to the recreation of an ancient tradition as to the creation of a new one. There was a premium on using models of the greatest antiquity, those which were the most 'pristine' (earliest, most primitive – with 'primitive' having positive connotations), a premium on going back to the 'purest founts'. The long-term effect of doing this turned out to be to produce 'progress', a going-forward, and ultimately to seeing this progress as a good thing. By looking backwards the men of what historians call the Renaissance developed for themselves a concept of progress and then sought to pursue it. In seeking to remake the past, they invented a new modern world.

In attempting to recreate ancient Rome and Athens anew in modern Italy, men began to look around for fresh relics of Antiquity. Writings of the Ancients that they had merely heard of but never seen, they began to find; they also found works of Ancients which they had never before known about. The wealth of Italian rulers, popes and patrons could fund expeditions to scour monasteries and the towns of the eastern Mediterranean. The collapse of the Greek empire of Byzantium and the fall of its capital Constantinople in mid-century (1453) had provided an astonishing windfall of Ancient works and wisdom. It also provided a fund of Greek refugees who were very willing to teach Greek to the Latins, which thus enabled them to tap the riches of ancient Greece for themselves. The Medicis of Florence played a particularly important role in this, especially with their support (1462–94) of the 'Academy' of Marsilio Ficino and Pico della Mirandola; some of the popes too, especially Nicholas V (pope 1447–55) and Leo X (pope 1513–21), put the resources of the Vatican to the support of artists and translators, and to the acquisition of books and manuscripts.

The men who promoted and led this campaign of renewal by following the example of the Ancients, generally answered to the name of 'humanist'. To mark their difference from their recent predecessors, they themselves coined a term to express the awfulness of the thousand-year gap between their models, the Ancients, and themselves, the loyal followers of these Ancients. The missing thousand years, the millenium without enlightenment, they called the 'Dark' or the Middle Ages. Coined as a term of disapprobation, the term 'Middle Ages' has entered the general vocabulary of historians. In their campaign, our humanists had at their disposal an instrument which was quite novel: the printing press. Printing by moveable types, invented around about 1450, enabled them to spread copies of the works of the Ancients abroad with an accuracy and a speed which had never before been known.

We shall look here at certain stages in the process of anatomy becoming a Renaissance discipline, taking three anatomists who published books in the new age of printing. They are, in order, Gabriele de Zerbis, who produced an enormously long *Book of the anatomy of the human body and of its individual members*, which appeared in print in 1502 in Venice; Alexander Benedictus (Alessandro Benedetti), who wrote a relatively brief *Anatomy, or five books on the history of the human body*, which was written probably *c.* 1493–97, and published in Venice in 1502; and Berengarius da Carpi (Berengario), who published his huge *Commentaries, with the fullest Additions, on the 'Anatomia' of Mundinus*, at Bologna in 1521.[1] All of these men came from and worked in Europe's richest and most intellectually lively area, north Italy, and at one time or another they were teachers at either Padua or Bologna, the two most important universities for medicine in the world at that time. Our discussion will indicate what they had in common in their anatomical projects; it will also show the disagreement among themselves over what the Ancients had been doing, and something of the way in which they projected back on the Ancients their own concerns; and it will show the early effect of the use of Greek on anatomizing . It will also draw attention to what we can regard as the first typically Renaissance emulation of an Ancient, though the person in question was not an Ancient at all but in fact a medieval. It was Mundinus.

The most fundamental thing that the anatomical projects of these three anatomists had in common, was that they were all following the Mundinus paradigm of seeing and demarcating the body. And, as the titles of de Zerbis's and Benedictus's books indicate, they take the *human* body as their subject, while Berengarius does so implicitly in his title by announcing that his book is based on that of Mundinus. All of these anatomists use the term 'anatomy' to cover how the body works as well as how it is structured (and how it is to be dissected). All of these works were meant, or could be used, to accompany a public dissection of the human body. De Zerbis, following the Mundinus order, divides his writing into *textus* sections to be read as the relevant part of the body is displayed in dissection, and into *additiones* which pursue matters further, and were designed to be read at leisure. Benedictus's book presents a fanciful demonstration he is giving to his friends, but it could certainly have been read to accompany a dissection if anyone had wanted to do so, and that demonstration would have been in the Mundinus order. Berengarius follows the Mundinus order remorselessly (though one doubts whether so much text could actually have been read at a public dissection before the body had completely rotted away), and the observables as we have met them in Mundinus are faithfully, indeed doggedly, followed through for every part. De Zerbis employs enormous typographical pointers for each of the observables, lest his reader be left in the slightest doubt (see Figure 3.1); Berengarius not only pursues the observables rigorously, but also points out where Mundinus was doing so.

Again, all these anatomists took the three venter division of the body for granted; it was simply something that they did not question. That Berengarius takes it for granted is implicit in the fact that he is commenting on Mundinus's text. De Zerbis, when he comes to deal with 'The anatomy of the middle venter in general, which is the chest (*pectus*)', after discussing the different ways of naming this region, writes:

> the dissector should proceed to showing the anatomy of the organic members of this venter by excoriating the middle venter beginning from below in the part which is called the 'ensiform';

Anathōīa vene artarialis

Paſſiones huic artarie eueniētes qualis ſint ē ſatis cōſpicabile:dicebat tñ Ariſto. 3°. de partib⁹. Circa artariā pulmonis perexigua accidere vitioꝝ genera:reſpectu.ſ.pulmonis: lyenis: ⁊ iecoris. de quibus ibidē ſermonē facit: Idq3 ratiōe recta. hec.n. maxime cordi aſſociatur. TEXTVS

Anathomia vene artarialis aut artarioſe:quā Hal.appellabat.6.de vti.par.vas artarioſum.

Subſtantia huius vene: quā diximus eſſe vnū vaſoꝝ duoruz intromittentiū māa in cor creata fuit geta.i. nō pulſatilis:⁊ in hoc ꝯmunicat alys ꝓprie vocatis venis: tamen illud in quo cōuenit cum artarys eſt:qꝛ duas hr̃e tunicas īuenta eſt: ⁊ ideo artariaz vocant venalē:dicebat Auic. Tū vt ſanguis ab ea effluens ⁊ reſudās per duas illius tunicas depurat⁹: colatus: ⁊ veluti cribratus ſit tenuiſſimus:ſiue in vltimo ſubtilitatis aliquātulū ſimilis ſubſtātie pulmonis:qm̄ paꝝ eſt ꝙ hic ſanguis exiuit a corde:eo ꝙ in via ⁊ trāſitu paruā morā duxit:⁊ etiā in corde:neq3 eſt ita decēter digeſtus ſicut ſanguis artarialis:quē cor per artariā venalē fundit in pulmonē.⁊ hoc eſt etiam quod dicebat Halyabb. ideo fuit duas habens tunicas:vt per eā qui ꝑuenit ſanguis ad pulmonez tenuis ſit ⁊ ſubtilis. ⁊ hic quo irrigat̃ pulmo ꝓpter ſui ſpongioſitatē corporis.omnia eniz mēbra ſibi cōfigurato ⁊ cōueniēti egent cibo. pulmo autez ſicut oſtēdemus ſubtilis eſt ſubſtantie:ideo ⁊ hꝰ eget cibo:Tum etiā vt ſanguis multū ⁊ diu ſiue perfecte digeratur in ea non reſudās cum abūdātia: ⁊ facil'r per duas tunicas donec ſit ſubtiliatus: Tuz etiā ſicut cōmuniter dicit̃.ideo tunicis duabus munita fuit ne motu ꝯtinuo cordis ⁊ pulmonis rumperetur.

Complexio hui⁹ vene eo redditur calidior complexiōe aliaꝝ venaꝝ:quo magis accedit ad naturā artarie:ꝓcipue tamen rōne ſanguis quē cōtinet ſubtiliorē ⁊ calidiorem.

Figura quātitas:⁊ numerus partiuz eius manifeſta fiēt in his que dicētur in anathōia venarum ⁊ artariaꝝ in genere.libro 3°.

Situm ⁊ colligantiā huius vene in ſua origine alio mō aſſignare iuenimus Galm̄:alio v̄o modo Auic.⁊ Halyabbatē.Gal.n.6.de vtili.parti. docebat venam artarialē explātari a duriori ⁊ dēſiori parte cordis dicēs. Neq3 eni ex ipſo corde:quāuis groſſum ſit ex alia parte ꝓdit artarioſum vas vel venoſuz:ſ3 ex ſubtilibus quidē ⁊ mollibus:ſimplicia ⁊ mollia ⁊ ſubtilia.duplicia v̄o rurſus ⁊ groſſa ⁊ dura ex ſtrictioribus.i. denſioribus eius explantantur particulis.

Auicꝰ v̄o ⁊ Halyabb.⁊ Auerroys p̄mo colliget.eaz docuerūt naſci a ramo maiori vene cōcaue aſcēdētis ⁊ implātate dextro cordis cōcauo:⁊ ꝯīter ab epate ſaltē mediate: Ita ꝙ ipſa oritur iuxta ſiniſtrā partez cordis ꝓpe locuz originis artariaꝝ.⁊ inde reflexa ad dextꝝ ventriculum trāſit ad pulmonē.modū autez aſſignāt dicētes: venā ꝑuenientē ab epate ad dextrā cordis aurē.dum eſt in directo cordis apud extenſionē ſiue ꝓtenſionē ei⁹ ad cor:vel emiſſionē ramoꝝ emittere a ſe tres venas:q̄rum p̄ma tēdit ad pulmonē:quā dicebat Auer. eſſe maiorē omniū aliaꝝ: ⁊ explātat̃ iuxta ſiniſtrā partē cordis ꝓpe locū originis artariarū:⁊ īde reflectitur ad dextruz ventriculū trāſiens ad pulmonē. Huius tamē cōtrariū per anathomiā īuentū eſt.nō enim cōſpicitur venaz ingredientē dextrā auriculā cordis vſq3 ad locuz originis artariaꝝ ꝓtendi.Seqtur ergo ex dictis eoruz venaz artarialē a corde originē ducere:⁊ nō a vena ītrāte auriculam cordis:nō enim eſt poſſibile venā artarialez a vena cōcaua:nec a ramo eius explātari ſm veritatē.⁊ propterea Raſis p̄mo ad Almāſorē nō numerat venā artarialē īter ramos vene cōcaue aſcēdentis. ADDITIO

Quod ſingulariter docebat Gal.6.de vtil.par.ca.13. his v̄bis.quoniā artarioſuz vas ex venoſo nō erat poſſibile explātari. oībus vtiq3 manifeſtuz eſt. vna .n.tunica eſt vene ⁊ ſubtilis nō vna:neq3 ſubtilis ea que artarie:ſed ſunt ei due tunice.Groſſa qdem ſufficiēter ⁊ dura:⁊ tēſa que eſt ab ītus ad inulos.i.villos diuiſa:trāſuerſos.mollis autē ⁊ ſubtilis ⁊ rara que ab extra. qualis ea que vene. Nō igit̃ erat poſſe ex vna ⁊ ſimplici tunica qualis eſt que cōcaue explātare duplicē groſſaz:ꝙ ſi forte Auic.⁊ Halyabbatē cōciliare libeat cuz Galō.cōſideranda ſunt verba Gal'i:cū dicebat.ex ſtrictiorib⁹ ⁊ dēſiorib⁹ particulis cordis:duplicia:⁊ groſſa: ⁊ dura vaſa explātari:que oīa recte cōſiderāda ſunt:⁊ hic eſſet opus ꝓlixiori ſermone. TEXTVS

Inueniūtur ꝓterea in orificio huius vene artarialis pāniculi tres qui ſine corde nō poſſunt illic firmari:propter qd̄ inſidēt cordi: atq3 in eo firmātur.quoꝝ iuuamentū eſt ꝓhibere ne hora ꝯſtrictionis pulmonis ſanguis delatus ad pulmonē ꝑ ipſam venā artarialē ad cor iterum recurrat:aut repellatur. ADDITIO

Sic enim docebat Gal. vbi ſupra dicēs. At vero neq3 mēbrana ſuperdēſari tā magna ⁊ talia qualia nunc ſūt in orificio artarioſe vene poſſibile erat abſq3 corde. ſed ēeni aliquā īfallacē ea habere debebāt in qua firmētur ⁊ aduaſtātur:vt firma ⁊ īflexa permanētia reſiſtant fluxibus materieꝝ ad id qd̄ retro quādo thorax agens fortiter:totū quidē pulmonē ītus colligit ⁊ contrahit cōꝓhēdēs in circuitu:ꝓmitq3 ⁊ ſtrigit vēas. TEXTVS

Eſt autem īceſſus trium hoſtioloꝝ huius vene artarialis ab interioribus cordis ad exterius.Cūq3 tendūtur ⁊ recte eleuātur fit ex eis vnuz coopertoriū claudēs totuz orificiū eius.ſ. vene artarialis:⁊ quādo ꝯſtringitur cor ⁊ ſanguis ab eo ad pulmonē impellitur per hāc venā expelluntur coopertoria quouſq3 applicētur vene. ADDITIO

Ita nāq3 docebat Gal. 5.de iuuamentis mēbrorum.cap.6.dicēs.Et qꝛ curſus iſtarum telaꝝ eſt ex īteriori cordis ad exterius eius:⁊ cōtinet os iſtius totius vene per hoc quādo expādūtur ⁊ eleuantur recte:fit ex oībus eoꝝ vnū coopertoriū magnum qd̄ claudit orificiū totū vene:⁊ quādo aliq̄ volunt exire ab īteriori cordis expellūtur coopertoria quouſq3 applicētur vene.⁊ ſic nō ꝓhibetur exitus rerum aliquaꝝ a corde:ſed per trāſitus illarum reꝝ erit velox leuis:quoniaz orificiū vene in illa hora ampliat̃ multū. TEXTVS

Iuuamentū ⁊ neceſſitas vene artarialis fuit vt pl'monē paſceret:dicebat Halyab. ADDITIO

Paſſiones huius vene ex dictis ⁊ dicēdis ī anathomia venaꝝ in genere ſunt ꝯſpicabiles:ei⁹ tñ opilatio ꝓſerti eſt timoroſa ꝓpter cām notā. TEXTVS

Anathomia pulmonis:q ⁊ pneumon grece dicitur.

Pulmonem nōnulli cordis ventilabrum appellāt: Nā eniz flabellū quoddā: ſicut adequatoriū cōſtituit pulmonis in loco pectoris caliditatē fontalē habēte:⁊ habere debente: vt poſſit alys in calore ſubuenire.Et ipſe quidē pulmo eſt mēbruz ſꝑmaticum īſtrumētale:⁊ viſcus alys viſcerib⁹ annumerādum mollis carnis vacu⁹ ⁊ īflatus:auctore Ariſto.p̄mo de hyſtorys.Mollis quidē carnis: vt obediētior eſſet dilatationi ⁊ cōſtrictioni.neq3 offendat mēbra:qb⁹ ipſe in ſuis

3.1 Textual pointers from Gabriele de Zerbi, *Liber Anathomiae Corporis Humani*, Venice, 1502, f. 72v, on the arterial vein (that is, the-vein-which-looks-like-an-artery).

> ascending upwards right to the top of the neck; for it is considered that spiritual members are contained right up to this point; that is to say, between the *fauces* [= top of the throat] at the higher part, and the *diaphragm* (which separates the spiritual parts from the natural parts) at the lower part.[2]

Benedictus writes similarly: 'the principal parts of the body are the head, the chest, and the stomach (*venter*)'.[3] When he turns to deal with the stomach he writes:

> The human body, which is to be dissected, is divided into three parts: a highest, a middle, and a lowest part. The first, the sensory part, contains the head. The middle is the spiritual part. The lower, in which the stomach lies, is called the natural part. In the first part the nobility of the senses has its place. In the second the source and cause of life (*vitae fons et ratio*) has its seat. The last is the workshop of Nature; from which part we need to begin, since the intestines rot more quickly . . .[4]

When he comes, in book 3, chapter 2, to the spiritual parts, Benedictus (using an argument of Aristotle on relative nobility) says they have their place

> between the head and the stomach; and the heart (for the sake of which all the spiritual members exist) is set in the middle of them since Nature is accustomed to put a more noble thing in more noble places: indeed higher rather than lower, and in the front part rather than the rear.[5]

Benedictus intends to name the parts in this venter: he will begin from

> that member by which the less noble parts are separated from the more noble, that is, the transverse septum [= diaphragm]. Lastly the members contained in the mouth itself (since they are dedicated to this part and to the stomach) will be spoken of in order.[6]

On the diaphragm we have a disagreement between our anatomists which is highly revealing for our purposes. For on the one hand Berengarius repeats the old charge that 'Aristotle did not know its use since, as Averroes says (*Colliget* 2, cxix), in the time of Aristotle the *scientia* of anatomy was not complete, as it was in the time of Galen'.[7] De Zerbis, on the other hand, defends Aristotle against this venerable charge of ignorance, maintaining that Aristotle did indeed know and refer to the diaphragm, though under another name.[8] But, as we have seen (above in Chapter 1) given what he was actually doing in anatomy, the question whether or not Aristotle 'knew about' the use of the diaphragm does not have much meaning, for Aristotle had been pursuing the anatomy not of man but of The Animal, and most animals do not have diaphragms because they are not mammals. For Galen, Averroes, Mundinus and our present anatomists, however, the diaphragm marked an important boundary in the body since they were pursuing the anatomy not of The Animal but of man (who *is* a mammal); thus it became an issue for them whether or not Aristotle had known about it. But this little disagreement is indicative of how these anatomists in general see Galen and Aristotle as having been engaged on the same general anatomical enterprise; the gross differences between the bodies they had seen are not yet visible.

It should perhaps be added that our three anatomists of course saw the heart, lung and trachea as a set of organs with a joint function; like Mundinus they regarded the (left) heart as a respiratory organ. Here for instance is Benedictus on the heart, lung and trachea as a set of organs with a joint, respiratory, function:

> ... through the hard artery [= trachea] *spiritus* is passed up and down from the mouth or nose ...[9]
>
> But we should move on to the lungs, which form the covering of the heart and temper the fervour of the heart with cold air ... The workshop of breathing has been placed in the lungs, continually attracting and returning the *anima* ... The lung changes the *spiritus* (as the liver changes the chyle) into the food of the vital spirit ... the lung encloses the heart in its hollow, and soothes it with a perpetual fanning ...[10]
>
> They say that the vital spirit dwells in the middle dwelling-place of the heart ... from this middle cavity the mouths of two vessels are seen to emerge, one of which (the larger) is joined to the aorta ... on the contraction of the heart this vessel dispenses the vital spirit to all the members through the arteries – right to the tips of the fingers ...[11]

The account is a little hard to follow in that Benedictus uses several terms for breath or air, some of which have more than one meaning. He is also talking (somewhat unusually for this date) in terms of a three-chambered heart. But there is no doubt about the way in which he sees the trachea-lung-heart as a set of organs united by function.

(ii)

De Zerbis (1445–1505) is our first publishing anatomist.[12] A son of Verona, he appears to have studied philosophy and medicine at Padua, and then stayed there to teach philosophy from 1467 until 1475. He then transferred to Bologna where he is recorded as having taught both medicine and logic over the years 1475 to 1483, rising to the highest position in philosophy as 'Ordinarius'. While here he published a commentary on Aristotle's *Metaphysics*, under the title *Quaestiones metaphysicae* in 1482. For nine years he was at the papal court in Rome, and then from 1494 until his death in 1505 he was Professor of Theory of Medicine at Padua. Like Bologna, Padua was a university of philosophy and medicine, and many students pursued a double degree in both subjects simultaneously. De Zerbis, as can be seen, pursued a teaching career in both subjects, and it is unsurprising that his approach to medicine (and hence to anatomy within it) was strongly influenced by his experience of teaching philosophy. De Zerbis gives his view of anatomy from the Theoretical chair. It is highly philosophical.

De Zerbis's knowledge of anatomy and medicine were not exclusively book-oriented. He appears to have performed dissections on his own account (for instance he refers to having dissected a pig in Rome); and he practised medicine and obtained a high reputation for doing so, as his presence at the papal court indicates. Indeed his skill in the practice of medicine was to lead him to treat the Turkish Pasha, and hence to suffer a horrific death, being sawn to death between two planks. But it is almost impossible not to notice that the exhaustive collocation of the doctrines of all possible authorities is indeed one of de Zerbis's main concerns. In this his is a typical 'scholastic' approach to anatomy. And it is the doctrine and truths of The Philosopher, Aristotle, which are the focus of de Zerbis's attention. Indeed on the title-page of his book de Zerbis calls himself 'Philosopher'. And, by contrast with some anatomists (such as Berengarius), de Zerbis treats anatomy as primarily a theoretical subject (a *scientia*); not as a practical, manual affair (an *ars*, or art).

What, for de Zerbis, was the point of pursuing anatomy? The primary point is encapsulated in the opening passage of the book:

> Natural things, and especially those which are endowed with vegetivity (*quae vegetantur*), have been established by the admirable intelligence of Nature: hence it is unworthy for anyone to have disdained the contemplation even of ignoble animate things, according to the authority of Aristotle. For there is nothing consonant with Nature in which it may be claimed that something admirable and divine has not been placed. It is said that Heraclitus (whom Lucretius calls 'The Obscure' and 'The Dark One') spoke to those people who, when they wanted to speak with him, hesitated to come close because they saw him sitting in a kind of hovel at the oven for the sake of the warmth: he ordered them to come in trustingly, 'for', he said, 'the immortal gods are not absent even from here, for all things are full of Jove'.
>
> This is why we should the more joyfully and accurately pursue the investigation of the human body and of its parts. And we are attracted the more eagerly to a consideration of the human body since in the construction and shaping of it not only has nothing been done at random or by chance, but most assuredly is the divine

> power (*numen*) of Nature more honorably and beautifully present in its construction than in all other animals: the capacities (*ingenium*) of man owe nothing to the mysteries in Samothrace and Eleusis (according to the authority of Galen) . . .[13]

And so on: the message is that in creating man Nature used her highest purposeful wisdom – investigating the body of man will reveal that wisdom to us. De Zerbis is paraphrasing two passages we have already encountered in the writings of the Ancients of anatomy: Aristotle in *On the parts of animals* (644 b 23– 645 a 24; see the text above, pp. 19–20), and Galen in *On the use of parts*, book 17 (see the text above, p. 30). Thus the point of pursuing anatomy was, for de Zerbis, primarily philosophical, by his understanding of the role of the philosopher: to reach and appreciate the wisdom of divine Nature, especially in forming man.

As a specifically Aristotelian philosopher, de Zerbis identifies with what he takes to be Aristotle's philosophical approach and doctrines in anatomy. Yet while Aristotle had been inviting people to the investigation of The Animal, de Zerbis reads this as an invitation to the investigation of man. De Zerbis is not guilty of misquotation; as a good Natural Philosopher of the schools he simply assumes that the high point of creation, man, was Aristotle's primary object of study in anatomy. To illustrate this concentration on the anatomy of man, here is a snatch of de Zerbis on the heart:

> *The anatomy of the heart: which the Greeks call cardia.* The heart viscus is the first root of all the viscera and members of the body; indeed it is, as it were, the source or spring of the innate and vital heat of life by which the animal is governed, according to the authority of Galen in the 6th book of *On the use of parts*.[14] The Peripatetics [i.e., the followers of Aristotle], dignifying it more highly, have taught that it has the same role in the human body as the First Intelligence does in the universe – from the fact that none of the animals endowed with blood (that is, blooded by their nature) lacks a heart, according to the authority of Aristotle in *On the parts of animals*, book 3;[15] indeed in the new-born every viscus appears more bloodlike and larger, and not without reason, for in the earliest stages of their formation it is a store and supply of material. There is neither a heart nor other viscus in any animal among the 'bloodless' creatures, yet in the 'blooded' ones all the viscera may be present . . .
>
> But leaving aside the heart of other animals, it is our concern here to deal only with the human heart. One should carefully observe that the articulation,[16] formation and organisation of the human heart excels the hearts of all other animals; for the hearts of those animals are more articulated which are better at sensing, but the hearts are less articulated in those animals which are more dull at sensing, such as the heart of pigs, according to Aristotle in book 3 of *On the parts of animals*.[17]

De Zerbis is using Aristotle's writings on animals as a resource for what he, de Zerbis, wants to say about man. And for his anatomy of man, the authority of Aristotle is the highest de Zerbis can imagine. For page after page de Zerbis continues his account of human anatomy, quoting and comparing all possible writers, but always giving the preference to Aristotle if he can. Completeness is one of the things de Zerbis is aiming at: completeness in the citation and critical comparison of authors and authorities. That is central to his research. His book is the product of many years work and (as de Zerbis intended) there is not a novel anatomical finding anywhere in it.

(iii)

Although he was an anatomist of this new age, this Renaissance, the pressure to learn Greek had hardly touched de Zerbis. The love of all things Greek did, however, greatly affect one of his contemporaries, who also published on anatomy and anatomizing: Alexander Benedictus (*c.* 1450–1512).[18] We shall use his case here to illustrate how two particular Renaissance things happened to anatomy: how it became 'Greek' again, and how it came to be performed in a 'theatre'.

Born in or near Verona, Benedictus seems to have acquired a degree in medicine from Padua around 1475. Over the period *c.* 1475 to 1490 he was practising medicine in Greece and Greek-speaking lands. Eventually the Venetian Senate invited him to become Ordinary Professor of Practical Medicine at Padua, where he remained until he died. He was thus a colleague of de Zerbis. His interests, his academic post, and his other publications (on the plague, and on diseases), all proclaim a practical physician. His mastery of Greek and his friendship with people such as Georgio Valla, proclaim a humanist.

What are the 'Greek' features of Benedictus's work? His *Anatomy* shows him in fighting mood on the side of the Greeks, and especially of Plato in the *Timaeus*. The dedication is addressed to the Holy Roman Emperor:

> I have decided to dedicate to you Maximilian Caesar Augustus a *Historia* of the human body, which the Greeks call 'Anatomicen', and we call 'the cutting of the members': an activity established by Philosophy for the use of healing, in which we observe the amazing and divine workmanship of God the Creator, that is, the body – in Plato's words, the sort of temporary vehicle of the soul; and in which we observe whence arise the felicitous objects of contemplation which the Greeks call *theoria*: images or phantasies, thoughts, memories and reason; we contemplate how the natural skill opened up the eyes, ears and other senses in the

> body, so that through those entrances *scientia* might come to the mind.
>
> The whole rationale of healing and practice, and the surgical discipline, depend on this too. When certain diseases invade one or several members, or if abscesses begin to grow, or an ulcer lurks within, or wounds are inflicted, or a member is dislocated or put out of joint, or something similar happens, nothing of these can be rightly understood without the dissection of the human body.

Thus for Benedictus anatomy is first and foremost a Greek subject, and a philosophical one; it is also crucial for anyone who wishes to be a skilled medical practitioner. But the aggressiveness of Benedictus's stance on behalf of the Greeks is shown in his attack on those of his contemporaries who continue to follow Arab influence and vocabulary in medicine:

> Amongst your laurel crowns, most learned prince, receive these little books on the *historia* of the members of man, which I have produced in Latin to the best of my ability, in which I have with much labour indicated many things which the more recent writers have omitted: those who have dishonoured the Latin language with foreign and divers errors – yea indeed with barbarous terms – and have inflicted a thousand wounds on the studies of medicine by their own ignorance; the younger generation has wretchedly inherited these terms in a great fog of things (*rerum*), and appears to be impelled headlong by the barbarian blast. Whence the very name of medicine has been for a long time transferred from the Greeks to the Arabs, with this consequence: that 'Arabian' audacity has grown to such an extent that they dare to fight openly with the Greeks over the credit for the origin of medicine . . .[19]

True to himself, Benedictus cites only Greeks (Plato, Aristotle, Hippocrates and Galen), writing indeed as if these Greeks were united in their view of the body. But although it is 'Greek' in its language and loyalty, Benedictus's little work is nevertheless squarely in the Mundinus tradition. We can see from this that keenness on and familiarity with the great Renaissance language, Greek, or keenness on Plato, the Ancient most celebrated in the Renaissance, was not in itself sufficient to get one to engage in a new anatomical project – and hence to allow one to see new things in the body or see the body in a new way.

But there was one important way in which Benedictus's work introduced something new, a new locus for anatomical demonstration: the Greek and Roman theatre. With Benedictus dissection was taken out of the lecture hall and into the theatre. It was in north Italy, near where Benedictus was living and writing, that deliberate efforts were being made at this very time by certain humanists to recreate Greek and Roman theatre as an event, and Roman theatres as places for it to be enacted in.[20] The Florentine humanist, Alberti, had engaged in a study of the ancient Roman architectural writer, Vitruvius, and this led to him producing his own work on architecture (1452, printed 1485) and designing a theatre in Rome for Pope Nicholas V in 1452. The work of Vitruvius, which deals quite extensively with designing theatres, was printed in 1486. The writings of the Roman playwrights Plautus, Terence and Seneca were put into print from the 1470s, while those of the Greek dramatists began to appear in print from the 1490s. For over a century, the new Renaissance reconstructions of Roman theatres were to be temporary wooden structures – just as the earliest ancient Roman ones had been – and were built either out of doors or within another building. They consisted of banks of tiered seating around a semi-circular 'orchestra' area, which faced an open stage (see Figures 3.2 and 3.3). The orchestra area (where the Greeks had played the drama) was reserved for seating the most important members of the audience. The first permanent theatres to be built were the Teatro Olimpico at Vicenza in 1585 (Figure 3.4), and the theatre at Sabbioneta in 1590.

A 'theatre' is, literally, 'a place for seeing': that is what the word basically means. It is in Benedictus's first little chapter that we find the earliest mention of the construction of an anatomical theatre. What he describes is a temporary structure, made of wood, erected specifically for an anatomical demonstration, and probably in the open air. The seats should be 'round a hollow'; thus, in shape the theatre should be like those in Rome or Verona, that is, an ellipse like the amphitheatres in those towns (not semi-circular like a Roman theatre proper). Benedictus had of course been familiar since his youth with the ruins of both the great Roman arena and the Roman theatre in his home town of Verona. For a century the anatomical theatre was to continue to be a temporary and seasonal structure, until the 1580s when, due to the local politics of Padua university, a permanent one was first set up.[21] Then, as the surviving building shows, it was still a wooden structure and still bearing very little relation to the shape of the building within which it was erected (see Figure 3.5). In such ways the anatomical theatre continued to be modelled on the secular theatre.

Benedictus begins the first chapter of his first book by recalling that in past times humans were vivisected (information he gets from Celsus, of course), and that kings and emperors themselves used to engage in this practice (which he gets from Pliny).[22] Today we do not do anything so barbarous, he says, but papal authority allows us to dissect dead criminals to investigate 'the intimate and arcane things of Nature'. The body of a hanged criminal, sufficiently fleshy to be visible to the spectators, is acquired; and then we are ready to start:

> For this a very cold winter is required, so that the cadavers do not immediately putrefy. Moreover, in a spacious ventilated place a temporary theatre is to be set up with seats around a hollow, like

3.2 A theatre: title-page woodcut of the comedies of Plautus, Venice, 1511 (*Ex emendationibus, adque commentariis Bernardi Saraceni, Ioannis Petri Vallae Plauti Comoediae XX recens singulari diligentia formalis excusae*).

3.3 A theatre: woodcut for the Lyons edition of the comedies of Terence, 1493.

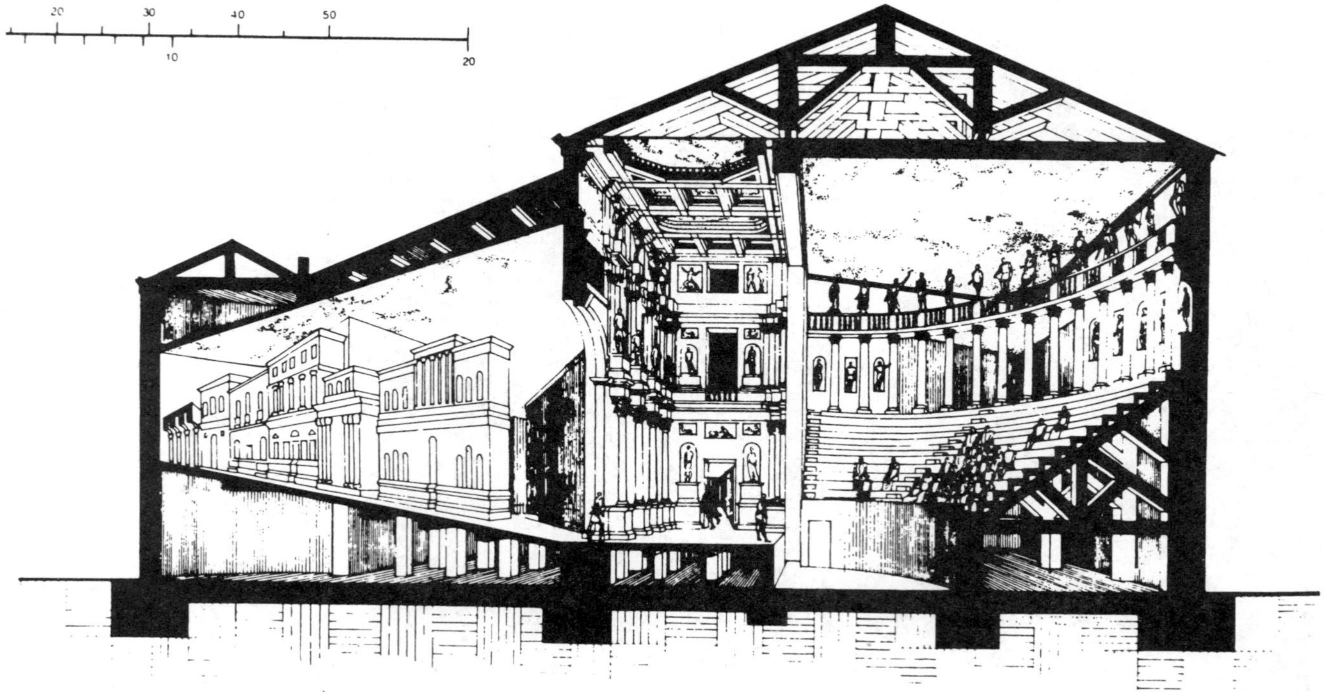

3.4 The Teatro Olimpico at Vicenza, 1585, one of the first permanent theatres to be built. Drawing by George C. Izenour.

3.5 The Anatomical Theatre of the University of Padua, 1594; scale model by E. Ashworth Underwood.

may be seen at Rome or Verona, of such a size that the number of spectators may be accommodated, and so that the surgeons (who are the dissectors) should not be inconvenienced by the crowd. These dissectors should be skillful: each of them should have dissected quite often before.

The order of seating is to be allocated according to the rank of the spectators. For this purpose there ought to be one *Praefectus* who takes care of and arranges everything. *Custodes* are to be appointed who should prevent the importunate plebs coming in. Two trustworthy *Questors* are selected, who should provide the

> necessaries out of the money collected. In this respect there is need of: razors, knives, hooks, drills, trepans (the Greeks call them *chenicia*), sponges with which the blood may be soaked up in the dissection, paring knives and bowls. Moreover torches ought to be got ready for the night. The cadaver is to be placed in the middle of the theatre, on an altar-like stage (*aeditiore scamno*), in a light place, convenient for the dissectors. A suitable duration of meeting is decided on, such that the task may be completely finished before the material putrifies.

The 35th (and last) little chapter of book five of Benedictus's book indicates the purpose of theatrical anatomizing for him. Here it is complete:

> *On the value of dissection*
>
> Now we will make an end to the dissection: the theatre is to be dismantled, since all the parts of the human body have been dealt with usefully rather than verbosely, and it will have been valuable to have run through the *historia* itself in a summary manner.
>
> I urge everyone, both beginners and veteran physicians and surgeons to attend a frequent 'theatre' of this kind, which should indeed be performed (*celebrandum sit*) every year, since in it we see the true, we contemplate the revealed.[23]
>
> In order that the works of Nature may fall under our eyes as if living, a text [e.g. on anatomy] is in general very similar to a picture which repeatedly rouses to action the negligence of recollection, and dispels the fog of the mind. But those people who, having trusted only in written accounts (as Plato says) without an inspection of 'things', are not turning over the impressed 'things themselves' in their minds; they are frequently deceived and consign opinion rather than truth to their minds – just like happens to those people newly beginning to navigate with maps in which the islands, bays, inlets and promontaries do not fully correspond to the true ones which fall under their eyes. But pictures of this kind do call to mind 'things' which have previously been seen with the eyes, and bring back as it were simulacra of the living things. I speak the living and animated speech of one who is knowledgeable (*scientis*), and I will call the *written* oration of it (not inappropriately) a kind of 'image'.

It is clear that for Benedictus the term 'theatre' has a double meaning: it is both the (temporary) physical place in which a special kind of event is shown, and it is the showing itself. What is physically on show in this theatre are the 'things themselves': in this case, the parts of the human body which are being laid open before our eyes. As they are laid open, the *historia* (the account) of them is related – what they are and why they are as they are – in the form of a brisk run-through of the observables.

But why should one attend this showing, and why attend repeated performances of it? Benedictus is writing at this point in terms of the

Aristotelian view of how the senses and the mind work. For present purposes the gist of it is this: if we have seen the 'things themselves' (*res ipsae*) with our eyes, then when we talk or reason about them, the mental 'images' we will be using will be ones which actually correspond to the 'things themselves', because they were produced in our minds by our looking at the 'things themselves'. Frequent attendances at these showings are necessary to keep these mental images stable and true to the 'things themselves'; an anatomy textbook will also help to do this – but only if we have actually seen the 'things themselves' with our eyes at least once. The relevance of Benedictus's attack on barbarous terms, and the grounds of his advocacy of proper Greek and Latin terminology should now be clear: we need to see the 'things themselves'; we need to be able to repeatedly stabilize or confirm the mental images or concepts of them; and we require the proper (i.e. the Greek or Latin) *words* for these *things*. Hence will be dispelled the great 'fog of things' created by the long use of barbarian terms; these barbarian terms simply confuse our mental grasp of the 'things themselves'.

It is most important to note the term that Benedictus uses to describe the performance that goes on in this anatomical theatre. It is *celebro*, which means both 'to perform' and 'to celebrate', as in a religious ritual. The primary purpose of the event is indeed religious; it is, as Benedictus says, to reveal 'the amazing and divine workmanship of God the Creator' in the highest point of His creation, the body of man. With the coming of the theatre, the sacred ritual of anatomizing is given its ritual space.

(iv)

We turn our attention now to the third of our early Renaissance anatomists: Berengarius da Carpi (*c.* 1465–1530), teaching at the old and eminent university of Bologna.[24] Berengarius was a surgeon by early training, and it was only after becoming skilled in this that he came to receive a literary education. At the age of about 19 he was educated for a while in the household of the signori of Carpi, under the tutelage of Aldus Manutius. This brought Berengarius to the fringes of the leading group of Medicean 'humanists' around Pico della Mirandola. From 1502 until 1526 Berengarius was Practicus of Surgery in the University of Bologna; he was quite a lowly figure in the medical faculty, and it was in this role that he was expected to teach anatomy. If we look at the woodcut on the title-page of Berengarius's Commentary, we can see an encapsulation of both his humanist and his surgical credentials (see Figure 3.6). At the foot of the woodcut on the title-page there is a

3.6 Title-page of Berengario da Carpi, *Commentaria ... super Anatomia Mundini*, Bologna, 1521.

picture of a dissection in progress. The professor sits in a chair (the cathedra), while some distance away a surgeon is portrayed performing the dissection of a human body. This little picture clearly puts Berengarius's teaching enterprise into the same format and tradition as that of Mundinus. And at the top of the page is the name of the Medici pope, Leo X, the great patron of humanist scholars, himself a product of the teaching of Marsilio Ficino and Pico della Mirandola, and who employed Berengarius on medical matters. The book's dedication is to another high-ranking Medici, Cardinal Guilio de' Medici, currently vice chancellor of the university and destined shortly to be pope himself, as Clement VII (1523–4). These put Berengarius's enterprise into the latest humanist tradition. However, no trace of concern with the recovery of Greek is visible in Berengarius's work. For Berengarius's aim is to reinstate, in a purer form, the work of Mundinus. Berengarius sees Mundinus as a true follower of Galen, and his aim (as the full title of his work declares) includes producing 'the text of the "Anatomia" of Mundinus restored to its pristine and true splendour'. So Berengarius is not simply working in the Mundinus paradigm, but seeking to restore the supposed pristine purity of that paradigm. Central to this endeavour was the re-establishing of the pure, 'pristine' text of Mundinus, that is, re-establishing the text as Mundinus intended it, free from the errors and accretions that generations of manuscript copyists had introduced into it; thus, it would be 'restored to its pristine and true splendour'.

Berengarius claims in his opening sentence that his intentions in doing this are to produce peace and concord among anatomists:

> When I saw so many and so great altercations between those writing on Anatomy, I resolved through means of a Commentary to draw together, by some quite brief summary, what I have seen by long experience in dissecting the bodies both of the living and the dead, and what I have sought in long reading. And my guide will be the excellent Mundinus of Bologna . . . In this exposition I will add some noteworthy things, not without their usefulness, from more recent writers, always with the senses and some authorities and reasonings of the divine Galen as my guide . . .[25]

The commentary will be his format in this venture, and this will allow him to discuss at length what Mundinus had in mind. He amply fulfils his promise.

Now, the comparison of authorities and hence (presumably) the bringing about of concord among anatomists, was also de Zerbis's aim; yet Berengarius frequently attacks de Zerbis. The reconciliation of doctrines that Berengarius has in mind must therefore be from a different direction. That direction is implicit in Berengarius's specification of the accomplishments requisite in a prospective anatomist, which reveals quite clearly what sort of activity he took anatomizing to be. It is highly manual and practical:

> I say that first and before all else it is essential that [the prospective anatomist] should be instructed in the *scientia* of medicine, together with its related subjects, and he should delight in reading – and in reading frequently – what Aristotle and Galen and Avicenna and others have said about animals, both living and dead, and what later writers have said; and he should be of good judgement so that he may make distinctions between the statements of the learned; and he should devote himself, to the best of his abilities, to the dissection of different species of animals both of diverse sexes and of diverse ages, pregnant and non-pregnant, alive and dead. And for acquiring this knowledge (*scientia*) he should devote himself to the anatomy of the fetuses of diverse species of animals, because many parts of the simple members of man may be very clearly seen through the anatomy of fetuses, parts which are not visible in the developed animals except in a lesser degree; and through the anatomy of such fetuses I acknowledge that I have learnt many things which were previously very obscure to me. One should not believe that it is possible to acquire this discipline through oral teaching alone or through books, since here sight and touch are required (as Galen says in book 12 of *On the use of parts*). Nor can a distinct knowledge of the members be acquired through just one anatomy, for when zeal is applied in the precise showing of one part, because of their closeness the composition of another part is almost completely destroyed. Many anatomies are therefore required . . .[26]

One reads the books of the authorities, but one should also acquire skill and practice in dissection: sight and touch are required. Who should be one's guide and model?

> One should put one's trust above all others in Galen, who has corrected the errors of other people, especially of Cleamenis, Praxagoras, and Philotimus, and other ancients (as Galen himself testifies in his book *On the anatomy of the uterus*): and common opinion holds that no error has yet been found in Galen's own anatomy. The opinions of Galen prevail firmly and inviolate to this day, although Avicenna in the books on animals holds, with Aristotle, an opinion against Galen on the origin of the nerves . . . Although both Aristotle and Avicenna dissent from Galen, yet they differ from him as philosophers. Galen, as a *sensibilis artifex*, does not go beyond the senses. In truth, Aristotle deals with anatomy somewhat 'in the universal', while Galen deals with it 'in the particular' and sufficiently completely – though Galen himself says that someone could come after him who would find things which he himself had not found. Yet the anatomist ought to trust Galen, especially when the senses agree with what Galen says: otherwise he should not; nor should he trust other authorities when experience and the senses are to the contrary . . . [27]

Berengarius here calls Galen a *sensibilis artifex*, a senses-oriented 'artist' (the implied dismissive contrast is with someone who deals in the theoretical, mental realm of the *scientiae*, the 'sciences'). It is clear that in Galen, Berengarius the surgeon-anatomist sees himself; that is why for Berengarius reliance on the senses, experience, observation, were characteristic of Galen's approach. The 'divine Galen' will be Berengarius's guide in finding out fresh knowledge: technical skill in dissecting will be the means. Hence it is going to be new knowledge in the tradition of Galen as Berengarius understood it: the Galen primarily of the dissection books (of whom Mundinus had been a loyal follower), rather than the more 'philosophical' Galen of *On the use of parts*. And that new knowledge will be about the body of man, primarily for the benefit of surgery and medicine.

Having characterized Galen in his own image, Berengarius could afford to notice and remark on certain differences between the (human) bodies he was dissecting and the (ape) bodies that Galen had dissected. Berengarius can notice, for instance, that in the human bodies he is dissecting the *rete mirabile* – crucial to Galen's account of the workings of man's body – is absent; and yet his faith in Galen as an anatomist is not shaken. (He reallocates the functions that Galen gave to the *rete mirabile* to the small arteries of the brain substance.)[28] He is not going to jettison Galen on the basis of a few mistakes committed because Galen was dissecting apes rather than man.

But, however inclined he was to the teachings of Galen the practical physician, and even though he was himself a practical surgeon dealing with medical problems every day, Berengarius did not think of anatomy as a discipline of exclusively medical use. When he discussed the 'utility' of anatomy, he wrote:

> the utility and necessity of anatomy needs to be known not only by the Physician, but also by the Philosopher exploring the secrets of Nature: here, through anatomy, Philosophers may very much marvel at the power of the Creator; as is clear from the 17th book *On the use of parts*, where Galen speaks about the Elephant and the Flea, and compares Nature to art, from the sculpted imitation of Phaethon with four horses most minutely and diligently sculpted: q.v.[29]

Exploring the secrets of Nature, admiring the power of the Creator, these are central goals of the anatomist, even of the surgeon-anatomist, the *sensibilis artifex* like Berengarius himself.

In the case of both de Zerbis and Berengarius, their research was pursued in the scholastic context; their anatomy books are typical products of the Schools in every respect. Berengarius, for instance, is writing a scholastic 'commentary', an exercise which (among other things) is designed to show the logical coherence and connection of the text under commentary. This was possibly the first time an anatomy text had been given this treatment.[30] In undertaking a scholastic project one takes it as given that the authors/authorities one is dealing with are correct in all essentials. The aim of scholastic exposition was not to find error in the authorities, nor (in the case of anatomy) to find discrepancies between the body and the book. Instead it was to find the concord which really unites the various authorities. If an authority seems to be wrong on any issue, then one reaches for ways of explaining this apparent aberration. Maybe some copyist made an error in transcription; maybe the author was using terms in a special way (which can be elicited by comparing all the instances of his use of this term in the text); maybe the author was speaking rather briefly, and the inherent truth of his statement requires subtle exposition to be revealed clearly. Thus it is reasoning, the deployment of linguistic and logical skills, which is the first resort for the scholastic exegete. Both de Zerbis and Berengarius employ all these techniques.

Sometimes, however, it appeared as if the doctrines or teachings of one authority really were at odds with the doctrines of another text-writing authority. The scholastic scholar took it as his duty to resolve these 'differences of doctrine' between one textual authority and another that his own scholastic activity had brought to light (and which in that sense was an artefact of his own activity). The doctrines of the authorities had to be 'saved', for otherwise they might cease to be authorities. For both de Zerbis and Berengarius, working in the early 1500s, it was a recurrent problem to reconcile what Aristotle and Galen and other authorities had said about certain parts of the body. We have already seen an instance of this with respect to the diaphragm. In their eyes Aristotle and Galen had sometimes disagreed with each other (they ignored the fact that Aristotle had been dead for centuries when Galen was writing, and were not aware that they had been describing different bodies). Such disagreement had to be resolved one way or another. If the methods described above failed, then there was one other usual way of deciding what the truth was, of deciding which authority was right. And this was to use prejudice: to give weight to one authority over another. De Zerbis, the self-styled Philosopher, usually prefers the opinion of Aristotle over that of Galen on such difficult occasions. Berengarius however gives his preference to 'the divine Galen'.

The approaches of both of these anatomists make assumptions about what the Ancients, and Aristotle and Galen in particular, had been up to. Naturally these assumptions match what that particular anatomist was himself doing in the early 1500s. De Zerbis takes anatomy to be pre-eminently a philosophical exercise. For him expertise in anatomy depends primarily on a good mind, not on skilful hands. This is why he usually sides with 'The Philosopher' Aristotle when there are matters in dispute. Berengarius, a trained surgeon, happiest when working with his hands and who came only late in life to a university post, and that post low in status, sides with 'The Physician' Galen. These allegiances to Aristotle or to Galen had thus been established for these anatomists by circumstances quite outside whether or not Aristotle's or Galen's doctrines about the body could be seen to correspond to the body. And by unconsciously assuming that the Ancients were engaged in the same anatomical activities that they themselves were, it never occurred to these anatomists that it might be an important question to ask what the anatomical projects of the Ancients had actually been.

(v)

Having seen some industrious anatomists of the early sixteenth century, all still working in the Mundinus tradition, we turn now to the next – and more typical or fulfilled – state of the Renaissance in anatomy. This is associated with the wholehearted embracing of Greek among anatomists, and with the rejection of 'medievals' such as Mundinus. This development was not accomplished without opposition, however.

The early decades of the sixteenth century witnessed great activity on the part of humanist physicians in recovering the texts and doctrines of the Ancients of medicine, and in this sense medicine too enjoyed as full a Renaissance as any other discipline. Galen's works were recovered virtually complete, and his anatomical writings and doctrines were thus recovered as part of this larger enterprise, and their initial treatment exactly paralleled that given to his works on medical practice.[31] This view, of the anatomical works of Galen being restored to practice in parallel with his medical works, is in contrast to the customary view of historians of anatomy that Galen's anatomical works were *rejected* at this period and replaced with something new and superior. So we begin by looking at the revival of Galen's greatest work on medicine, and then turn to seeing the parallels between this and the revival of his anatomical writings.

On the practice of medicine Galen had written much: one enormous book, the *Method of healing*, and several smaller ones. The *Method of healing* had been available in Latin hitherto in a compressed and in a full version from the Arabic, and in a partial version from the Greek.[32] A full Greek manuscript of it was acquired by the Ferrara physician Nicolaus Leonicenus. He lent it out strictly on the condition that it should be printed, and the Greek text thus appeared in 1500 at Venice.

Thomas Linacre, a member of the 'Academy' of the Venetian printer Aldus, and who had been taught Greek at Padua by Chalcondylas and others, translated it complete into classical Latin for the first (and last) time and had it published in Paris in 1519. This new jewel, the *Method of healing*, became the centre of the attention of the man who was in turn professor of the Practice and of the Theory of Medicine at Padua, Joannes Baptista Montanus (1498–1552), and it was he who most thoroughly explored its potential as a new medical gospel. But because Montanus was tardy at publishing his own teachings, it was left to his pupils of the 1540s and 1550s (excited beyond all measure at the power and potential of the message of the book and at Montanus's exposition of it) to do so on his behalf. Johannes Crato von Crafftheim, Bassianus Landus (and through them Theodor Zwinger, Julius Alexandrinus and Hieronymus Mercurialis), and John Caius, travelling back to their homes after being students at Padua, each produced books which announced this astonishing and powerful method of healing to the world.[33] The version produced by Caius, for instance, published in Latin, is called *Two books on the method of healing, according to the sentiment of the most distinguished physicians Cl. Galen of Pergamum and Jo. Baptist Montanus of Verona* (Basle, 1544).[34] As faithfully as he could, Caius was recapturing and recounting the teachings of his two great masters, ancient and modern. What was the great new message? That the physician could and should diagnose disease by starting from the signs and symptoms exhibited by a particular patient, and thence following a sequence of strict reasoning which would lead him to understand the internal – the hidden – *cause* of the disease; and from knowing the cause, the physician would then be able to reason his way to finding, preparing and administering the proper *cure* and adjusting it to suit his particular patient. Whether a particular disease could be cured, and (if so) how, could be worked out by mastering and strictly following a logical procedure. This was the method, and it was universally applicable. Thus although Galen's writings on diagnosis, cause and cure had, in one form or another been the basis of medicine for centuries, it was also something new and exciting – and it seemed to open the door to a future of better and more successful medical practice – to follow these writings properly and fully.

Thus although Galen's doctrines were centuries old, yet a whole generation of men were now dedicating themselves anew to following the master, the Prince of Physicians, and setting out to practise, as authentically as they could, the medicine of Galen. They believed they were the first to do so since Galen himself. Physicians of mid-century thus proselytized energetically about the supreme value and importance of Galen's doctrines. They were engaged in an active campaign, because they felt they had to win over their fellow physicians to a truly Galenic mode of practice. One of the things they felt compelled to do was to prepare new compact guides to the Galenic way of understanding and curing the human body. Leonhard Fuchs in Tübingen in the 1530s and 1540s, Jean Fernel in Paris in the 1540s and 1550s, produced some of the earliest such *Institutions*, *Compendia* and *Epitomes*, and they were to be followed by dozens of other teachers over the next century.

In all this we need to appreciate that the aim of these practical physicians was not primarily to assess, judge or argue with Galen's doctrines, but to master and practise what he had taught; for this (they believed) would in itself lead to the improvement of medical practice and restore it to the standards of Galen himself. They were trying to do what Galen said a physician should do. But we also need to remember that these physicians lived in a world where 'official' medical practice was now divided among three groups of practitioners: apothecaries who made up drugs, surgeons who operated on the outside of the body, and themselves (the physicians) whose self-chosen province was *internal* medicine. These groups had (and the physicians believed that they should continue to have) different kinds of education, different rewards, and different statuses – their own being of course the highest. Galen himself, by contrast, had had a range of practice which had no room for these distinctions. Unknown to themselves, and despite their best efforts and intentions, they were thus promoting a form of practice which was not like that of Galen himself.

The same was true about anatomy, the subject which Galen had regarded as underlying all proper medicine. In the 1530s one of the centres of activity in translating and synthesizing the anatomical works of Galen was Paris. When the French had descended on north Italy in 1494, they had brought back many things to Paris, including a thirst for the teaching of Greek on the part of some Parisian scholars, which was soon satisfied by the arrival of Greek refugee teachers. In the first two decades of the century there was really only one person in the Paris medical faculty yet able to translate Galen, Gulielmus Copus (Wilhelm Kopp, or Cop) from Basel, though several of Galen's works were printed at Paris printing shops. But in the late 1520s Cop was joined by other enthusiasts, and especially by Johann Guenther (Johannes Winther, 1487–1574) from Andernach near Cologne, and Jacobus Sylvius (Jacques Dubois, 1478–1555) of Amiens. Both were active members of the medical faculty at Paris university; both engaged in the work of translation of Galen's anatomical works; both presided at public dissections, and also at private ones; both produced compact summaries of Galen's anatomical doctrines. Together they played a major role in making the anatomical Galen available and in urging others to restore anatomy.

Guenther had been educated at Utrecht and other places, where he had learnt Greek from a young age. In the early 1520s he was at the 'three language college' at Louvain, teaching Greek. By now he was about 37 years old, but nevertheless decided it was not too late to study medicine, and for this he went to Paris where he followed the lectures available in the Medical Faculty. In 1528 he was made Bachelor, and in 1530 Doctor. He was even appointed one of the Royal Physicians in 1535. Sylvius was also a late qualifier. He did not take his medical degrees until 1529, when he was over 50. Like Guenther, Sylvius devoted the years of his maturity in Paris to publishing works promoting and expounding the writings of Galen. He was even appointed a professor at the new humanist Collège Royal.

First let us deal with their translations of Galen's anatomical works. Guenther translated *On anatomical administrations* into Latin, and published this in 1531 at Paris, and it was printed again in Basle the same year.[35] It immediately dominated the field. Sylvius, together with Martinus Gregorius (Grégoire), issued a revised edition of Galen's enormous *On the use of parts*; it was still Niccolo da Reggio's Latin translation of 1317–22 but, as Sylvius and his collaborator say on the title-page, 'corrected for a third time with greater care according to the truth of a Greek exemplar'.[36] It should be noted, to Sylvius's credit, that this was the nearest this huge book ever got to a new humanist translation. Between them, therefore, these two friends, Guenther and Sylvius, were responsible for the appearance at Paris of Galen's two central anatomical texts, translated or corrected anew from the Greek: the work which told one how to dissect, and the work which told one how to use anatomy to reveal, understand and admire the purpose of Nature in forming the parts of the body.

But they did not stop here. Both of them also wrote and published small guides or summaries to Galen's anatomical doctrines. That of Sylvius was not published until 1555, after his death and after much had happened in the world of anatomizing, and we shall return to it in a later chapter.[37] But Guenther's little book came out in 1536 and was called *Four Books of Anatomical Institutions for Students of Medicine, According to the Opinions of Galen.*[38] In his Preface Guenther addresses his dedicatee and says,

> When I was recently publicly expounding the *historia* of the human body to students, it occurred to me that it would be valuable to arrange in a commentary the things which became visible to us in dissection – a commentary which might be like a compendium of all the things which Galen, the Prince of Medicine, has dealt with more fully and in many books. I have done this the more readily for this reason, that there is no recent writer whom we may suitably and safely take as a pattern; and also because the knowledge

> of Anatomy, useful and necessary not only to physicians and philosophers but also beautiful and honourable to all kinds of men, today lies utterly neglected.[39]

This was not always the case, for kings and emperors once cultivated anatomy.[40] After deploring the ignorance of many people who claim to be physicians yet know no anatomy, Guenther explains the nature of his book:

> So that I might perform a favour to those learning medicine properly, I have contracted briefly into four books the things which up to now I have found to be true from Galen in his own writings, using that order and rationale of teaching which I see to have been approved by Galen and hope will be most easy for students of Medicine.
>
> First I expound the site of each part, its number, the distinctive nature of its substance, its magnitude, figure and connection: then I add the use of each part, and the means of dissecting it: helped by these, any reasonably intelligent person will be able to do the dissection himself, without anyone-else's help.
>
> Now, in dissecting I have followed the ancient custom of scholars, beginning from the lower venter of the body – and notwithstanding that Galen used to dissect the muscles of the abdomen first for his friends – then the intestines, stomach, liver, spleen, kidneys, and bladder (if it was a woman, also the womb), lest these, left till later, should putrefy; then I have dealt with the useful part of Anatomy, which we see to be neglected not only by physicians but also by surgeons.
>
> In the second book I expound the structure of the thorax, and its parts both internal and external. In the third I show the nature of the skull and its processes. In the fourth I deal with the dissection of the limbs, which first comes to be understood after the knowledge of the bones. This most extremely useful part I pursue more diligently in this book, because none of the recent writers have left anything in writing about it.[41]

What relationship does Guenther's work here bear to that of Galen? 'As for my contribution', Guenther writes,

> there is nothing written here which is not redolent of Galen, whose doctrine I deliberately follow: there is nothing asserted which I have not seen him write.
>
> But if anyone holds the work in esteem by the name of Galen, because all the things are Galen's and described virtually word-for-word from him, to him would I say that the work is not Galen's but mine. Equally, if some other person should despise my work, he should know that all the things are Galen's, and if he thinks my things are to be despised, he is actually despising Galen.[42]

It can be seen that Guenther's anatomical project is still firmly in the Mundinus paradigm, even though Guenther has a much wider knowledge of Galen's anatomical writings – and in Greek at that – than anyone of an earlier generation. Books 1, 2 and 3 respectively deal with the lower, middle and highest venters. The observables are dealt with in each case. The anatomizing was performed on a dead human. The novelty of Guenther's approach, as he himself saw it, was book 4, for here he dealt with the anatomy of the limbs, something neglected in all medical schools (he says)[43] and yet most important for the cure of injury; he deals with the bones and muscles, and then with the arteries, veins and nerves of the limbs. Many years later, after many important and radical changes had occurred in anatomizing, Guenther was still recommending that anatomy should be learnt not only from listening to lecturers and reading books, but also from the inspection of bodies and visual evidence (*ex fide oculata*),[44] and that the most appropriate sequence was to deal with the bones, then the three venters.[45] He also recalled that at Paris over 30 years earlier (that is, in the late 1530s) he had anatomized in the sequence (i) the skeleton, (ii) the three venters, and then (iii), using other bodies, demonstrated the muscles and nerves, and the veins and arteries; and that subsequent famous anatomists had copied this sequence from him.[46]

Guenther's anatomical work of the 1520s and 1530s, like that of Sylvius, thus has four aspects: they *translate* Galen from the Greek, they *dissect* to see what Galen was talking about, they *reduce* Galen's doctrines into handy compendia, and they *publish* them in order to spread the good word to a wide audience. And they set out to follow Galen's doctrine in every respect: 'there is nothing written here which is not redolent of Galen, whose doctrine I deliberately follow', as Guenther wrote. And when Galen did later come under attack, Sylvius distinguished himself as the man who vindicated Galen from his calumniators.[47] For Guenther and Sylvius the right route was to follow the *doctrines* of Galen.

Guenther, however, had to leave Paris in the 1540s when he became infected by the new faith, Protestantism. Eventually, in the 1570s, he was to produce a massive work on learning medicine whose title reveals his willingness – though a convert to the new – to pursue a middle path between classical and modern medicine: *Two Commentaries on both Learning and Practising Old and New Medicine*.

Notes

1. The editions I have used are, *Liber Anathomiae Corporis Humani et Singulorum Membrorum Illius: Editus Per Excellentissimum Philosophum ac Medicum D. Gabrielem de Zerbis Vernonenses*, Venice, Locatellus, 1502; *Alexandri Benedicti Medici Clariss. Anatomice, sive De Historia Corporis Humani, Libri Quinque*, Strasbourg, 1528 (first published 1502);

and *Carpi Commentaria cum Amplissimis Additionibus Super Anatomia Mundini Una Cum Textu Eiusdem in Pristinum et Verum Nitorem Redacto*, Bologna, de Benedictis, 1521.

2. de Zerbis, op.cit., f. 52 v. My translation.
3. P. 10v. My translation.
4. P. 8r.
5. P. 36v (the Aristotle passage is 665 b 20–5).
6. P. 37r.
7. Berengarius, op.cit., f. cccxxvii r.
8. de Zerbis, op.cit., f. 59v.
9. P. 40 v.
10. P. 39 v.
11. P. 45 r and v.
12. For the life of de Zerbis I rely on L.R. Lind, *Pre-Vesalian Anatomy: Biography, Translations, Documents*, Philadelphia, American Philosophical Society, 1975, pp. 141–56.
13. de Zerbis, op.cit., folio 2r.
14. See the translation by M.T. May, *Galen on the Usefulness of the Parts of the Body*, Cornell University Press, 2 vols, 1968, vol. 1, p. 292.
15. Aristotle 665 a 29 – b 20.
16. I.e. the structures permitting flexibility, akin to the sutures in the skull.
17. de Zerbis, op.cit., folio 62r, my translation. The passage from Aristotle he is referring to is 667 a 7–10. Compare the translation of this passage from Zerbi offered by Lind, op.cit., p. 155.
18. For biographical information on Benedictus I rely on Lind, *Studies in Pre-Vesalian Anatomy*, pp. 69–137. Translations are my own.
19. P. 7r.
20. On the revival of the theatre, and especially on the physical structures, see the entry 'Theatres and Stages' by George C. Izenour, in *Encyclopaedia Britannica* (1979 edition); the entry 'Architecture' by Victor Glasstone in Phyllis Hartnoll *The Oxford Companion to the Theatre*, Oxford University Press, 3rd edition, 1967; Frances Yates, *Theatre of the World*, London, Routledge and Kegan Paul, 1978 (first published 1969), especially chapter VII; Glynne Wickham, *A History of the Theatre*, Oxford, Phaidon, 1985.
21. On the general history of anatomy theatres, see Giovanna Ferrari, 'Public Anatomy Lessons and the Carnival: The Anatomy Theatre of Bologna', *Past and Present* no. 117, 1987, pp. 50–106; on pp. 84–86 she deals particularly with the relation of theatre design to anatomical theatre arrangements, quoting Estienne, who wrote in 1530 that 'the anatomy table should be arranged in front of the theatre, in the place where the ancients placed the stage'. The Ancients here must be the Romans, as the Greeks used the orchestra area for acting. The history of the creation of the first permanent anatomical theatre at Padua in the 1580s and 1590s will be transformed by Richard Palmer's forthcoming paper in *History of Universities*. See also Arturo Castiglioni, 'The Origin and Development of the Anatomical Theater to the End of the Renaissance' *Ciba Symposium*, 1941, pp. 826–844; Gottfried Richter, *Das Anatomische Theater*, Abhandlungen zur Geschichte der Medizin und der Naturwissenschaften, Heft 16, Berlin, Ebering, 1936, pp. 30–34.
22. See above, Chapter 1, note 23.

23. I have chosen to break Benedictus's very long sentence at this point.
24. On Berengarius the authority is now R.K. French, 'Berengario da Carpi and the use of commentary in anatomical teaching', in A. Wear, R.K. French and I.M. Lonie, *The Medical Renaissance of the Sixteenth Century*, Cambridge University Press, 1985, pp. 42–74.
25. Berengarius, op.cit., folio iiii r; my translation.
26. Berengarius, op.cit., folio vi verso.
27. Ibid., folio vi v. to vii r. (this passage is continuous with the previous one quoted).
28. See Edwin Clarke and C.D. O'Malley, *The Human Brain and Spinal Cord. A Historical Study Illustrated by Writings from Antiquity to the Twentieth Century*, Berkeley, University of California Press, 1968, pp. 764–66, where the pertinent passages are translated.
29. Berengarius, op. cit., folio v r.
30. French, op.cit., . 43.
31. On the editions of Galen see the indispensable work of Richard J. Durling, 'A Chronological Census of Renaissance Editions and Translations of Galen', *Journal of the Warburg and Courtauld Institutes*, 1961, 24, pp. 230–305. On the general issues of Greek translations of classical medical works, see Vivian Nutton, 'Greek Science in the Sixteenth Century Renaissance', in J.V. Field and Frank A.J.L. James, *Renaissance and Revolution: Humanists, Scholars, Craftsmen and Natural Philosophers in Early Modern Europe*, Cambridge University Press, 1993, pp. 15–28; however, Nutton does not deal with anatomy in this article.
32. See Pearl Kibre and Irving A. Kelter, 'Galen's *Methodus Medendi* in the Middle Ages', *History and Philosophy of the Life Sciences*, 1987, 9, pp. 17–36.
33. On these people see Johannes Karcher, *Theodor Zwinger und seine Zeitgenossen*, (Studien zur Geschichte der Wissenschaften in Basel, III), Basel, Helbing und Lichtenhahn, 1956.
34. Reprinted in E.S. Roberts (ed.), *The Works of John Caius*, M.D., Cambridge University Press, 1912.
35. *De Anatomicis Administrationibus Libri Novem, Joanne Guinterio Andernaco Interprete*, Paris, Colinaeus, 1531; for the Basle edition of 1531, see Durling, op.cit.
36. I have used the Paris 1543 edition; the first edition was published in 1528.
37. *In Hippocratis et Galeni Physiologiae Partem Anatomicam Isagoge*; I have used the Venice 1556 edition; the first edition was Paris 1555. On Sylvius see C.E. Kellett, 'Sylvius and the Reform of Anatomy', *Medical History*, 1961, 5, 101–16; and Gerhard Baader, 'Jacques Dubois as a Practitioner', in A. Wear, R.K. French and I.M. Lonie, *The Medical Renaissance of the Sixteenth Century*, Cambridge University Press, 1985, pp. 146–154.
38. *Institutionum Anatomicarum Secundum Galeni Sententiam, ad Candidatos Medicinae, Libri Quatuor*, Paris, Colines, 1536. On the editions see Josiah C. Trent, 'The *Editio Princeps* of Guenther's *Institutiones Anatomicae*', *Bulletin of the History of Medicine*, 1945, 18, pp. 109–11. On Guenther, see Louis-Antoine-Prosper Hérissant, *Eloge Historique de J. Gonthier d'Andernach*, Paris, J-T. Hérissant, 1765.
39. *Institutionum . . . Libri Quatuor*, f. Aii recto-verso. My translation.

40. See above, Chapter 1, note 23.
41. Ibid., folio Avi recto.
42. Ibid, folio Avii recto.
43. Ibid., p. 111.
44. *De Medicina Veteri et Nova tum Cognoscenda tum Facienda Commentarii Duo*, Basle, Ex Officina Henricpetrina, 1571, Comm. I, p. 91.
45. Ibid., p. 139.
46. Ibid., p. 91.
47. Sylvius, op.cit., publisher's preface, p. 3.

CHAPTER FOUR

Vesalius: The Revival of Galenic Anatomy

We now turn to the most famous of all sixteenth-century anatomists, Vesalius (Andreas Van Wesele, 1514–1564). Vesalius represents a further stage in the revival of Galen's anatomy in that he taught the world to see a different Galenic body; and he taught anatomists, physicians and philosophers to adopt a new ambition with respect to the Ancients of anatomy. Hitherto the achievement of Vesalius has been judged by historians to be the creation or introduction of 'scientific' anatomy (or of 'modern' anatomy, or of introducing 'the scientific principle' into anatomy), and it has been taken for granted that this necessarily and laudably involved the repudiation of Galen and all he stood for; to become modern and scientific, anatomy (in the historians' eyes) had to be 'liberated' from the authority of Galen, and to be based on experience and observation. The historians of Vesalius have then read back this 'achievement' into his intentions. That is to say it has been implicit or explicit in virtually all earlier accounts of Vesalius (at least all those that I know about), that he *set out to* repudiate, correct, or replace Galen.[1] We shall, however, find his achievement (and also his intentions) to have been quite otherwise. It will be necessary to bear in mind that although Vesalius produced the largest and most beautiful and faithful illustrations of the dissected human body yet seen, and in this sense can be seen as associated with the Renaissance in the visual arts, yet he was not the first to think visually about the body nor the first to put illustrations of it into print: a number of his contemporaries were engaged in doing just this. The novelty of what Vesalius did lay elsewhere. To appreciate what it was we need to trace his activity in the frantic years of his twenties.

(i)

Vesalius, a whirlwind of a man, came out of the north. Not out of the dark and unenlightened north, but from Italy's 'northern depot' (as it has been called), Flanders.[2] His family were Flemings. Vesalius himself was born at Brussels. His father was apothecary to the Holy Roman Emperor, Charles V. His grandfather had, however, been a physician to the Emperor Maximilian, and his great-grandfather had been city physician to the city of Brussels. Why Vesalius's father had not followed them into medicine proper would seem to have had much to do with the fact that he was illegitimate (being legitimized by imperial favour only in 1531).[3] While Vesalius was an infant, his father was with the imperial court, which at that time was in Spain. Part of Vesalius's education seems to have been at the new humanist tri-lingual college at Louvain (1530–), where he learnt Greek and Hebrew and the best humanist Latin.

In 1533, intent on pursuing medicine, Vesalius went to Paris and registered as a student in the medical faculty. Vesalius found the Parisian physicians in the forefront of reforming medicine on the Greek model. As he shortly afterwards recalled, the principles of combating disease in use among most physicians differed greatly from the practice of the ancient Greeks, but instead followed the Arabs. The best thing, he thought, was to compare the writings of the Arabs with those of the Greeks to get the best of both worlds. And, he claimed, this procedure was particularly recommended by 'the excellent Parisian physicians to their students, in the hope that soon the ancient art of medicine starting from Hippocrates will be, to the great advantage of mortals, recalled to memory and will come to flourish, through the gradually increased labours of those who practise this art'. This sequence would take a student through the medical art from the more general to the more particular, concluding at the curative part.[4]

At Paris Vesalius also came under the teaching of Sylvius and Guenther, and assisted them in their dissecting and textual work on Galen's anatomical legacy. When Vesalius met them they were working to fully grasp the doctrines of the great medical Ancient, Galen, and were engaged in producing editions of his writings and systematizing the truths of them into compendia. They were also proclaiming the new message that anatomy is the very basis of medicine, and hence ought to be treated as fundamental to all other medical study. Vesalius himself became involved with them in their work on this, the most fundamental of all the parts of medical knowledge. In these circumstances his initial assumption that Galen had been dissecting human bodies (though at odds with what Galen himself says) is quite understandable. His abounding natural talent showed early: Guenther in his own *Anatomical Institutions* of 1536 says that he discovered the course of the spermatic vessels through 'the help of Andreas Vesalius, son of the Emperor's apothecary, a young man (by Hercules) of great promise and, in addition to an extraordinary knowledge of medicine, learned in both languages [that is, Greek and Latin], and very skilled in the dissection of bodies'[5], which is quite a compliment to the dissecting skills of a young man of 20. For in Paris, as he became acquainted with the anatomical writings of Galen and sought in the company of his teachers and fellow-students to locate and trace out these truths on the body, Vesalius discovered within himself a total passion for dissecting, and he quickly acquired a skill in dissecting that exceeded that of any of his teachers. Indeed, it would be no exaggeration to say that Vesalius rapidly became the most skilled, elegant and thorough dissector since Galen himself. He was the best dissector there had been for thirteen centuries. And he knew it.

In Paris Vesalius had access to the bodies legally provided to the medical faculty. He also went hunting for human bones in the Paris cemeteries, and he cut up live dogs too. At the outbreak of war in France (1535–36), Vesalius left Paris without a medical degree, and returned to Louvain. Here he continued his impassioned search for human bones, on one occasion stealing and reassembling the skeleton of a hanged man. He attended as many autopsies as he could; and he performed anatomies in the medical faculty. After taking his bachelor's degree in medicine at Louvain in 1537, where for his thesis he defended a paraphrase of a practical work by the Arab physician Rhazes, Vesalius went to 'the true nurse of talents' (as he called it),[6] Italy, and to the world's most famous university for medicine, Padua. In Padua Vesalius was considered to be a northerner, a 'German'. The attraction of Padua for students from outside the Veneto was of long standing, and there were crowds of such students. The *lingua franca* of the students was of course Latin, in which all teaching was conducted.

Once arrived at Padua, Vesalius took the (oral) examinations for the degree of doctor of medicine as soon as possible. The heroic myth has it that the very next day Vesalius was appointed 'Professor of surgery and anatomy'. The truth is a little less exciting: he seems to have been appointed as the demonstrator of anatomy and to give the lectures on surgery.[7] There are no records extant of his appointment, and all that we can feel sure of is that Vesalius would have actively sought this post, for it gave him licence to do the thing that he most loved and at which he excelled. But if the position of demonstrator of anatomy was lowly when Vesalius took it, it was his conduct in this position which was to lead directly to a great rise in the status of the post to that of *Professor* of Anatomy, both in Padua and in every other self-respecting medical faculty. Within a few days he was able to demonstrate his unrivalled skills in dissection at an anatomy (6–24 December 1537) on the corpse of an 18-year-old male. Notes on this anatomy survive from a student friend of Vesalius.[8] He performed the dissection, we should note, in the Mundinus order; and, if the statutes are anything to go by, he did it to accompany the reading of the text of Mundinus by a senior professor.

Already, in his first surgical lectures and his first anatomical demonstration, Vesalius was thinking visually. These are the sketches in the student's notes, copied down from diagrams produced by Vesalius in course of the anatomical demonstrations (see Figures 4.1 and 4.2). Of the distribution of the veins, arteries and nerves, the student noted: 'since the memory of this sight may be confirmed from the illustration which the Surgeon gave, and which he intends to publish, I omit it here'.[9]

We can see from this that Vesalius's scheme of publishing illustrations of the dissected body dates from very early in his career. His own account of how it all started declares that in one of his early surgical lectures at Padua he was dealing with the cure of inflammation; intending to expound the thoughts of the divine Hippocrates and of Galen on revulsion and derivation (that is, on where to bleed to cure diseases), he drew the veins on a piece of paper, thinking that it would thus be easy to demonstrate what Hippocrates had meant by the term *kat' ixin*, 'in a straight line'. This delineation of the veins was so pleasing to all the professors and students of medicine (Vesalius claimed) that they strenu-

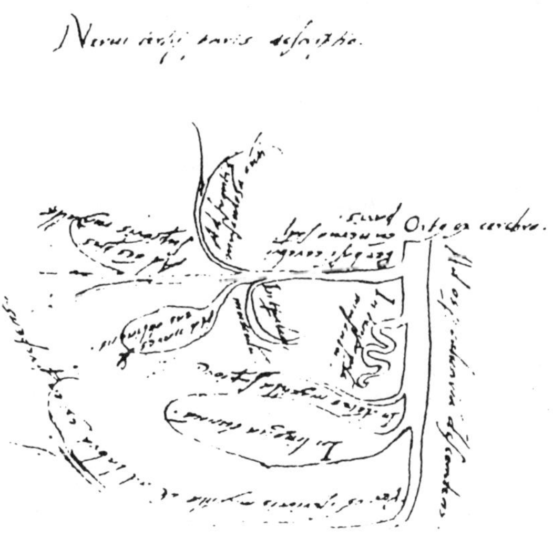

4.1 Sketch of the distribution of the third pair of nerves, by Vitus Tritonicus, copied from Vesalius's own sketch in his first Padua demonstration, 6–24 December 1537.

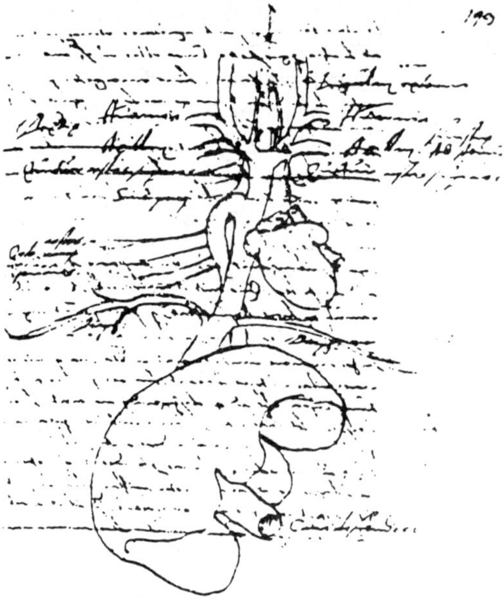

4.2 Sketch of the liver and the vena cava, as the source and distribution of the veins, by Vitus Tritonicus, copied from Vesalius's own sketch in his first Padua demonstration, 6–24 December 1537.

ously entreated from him a similar description of the arteries and of the nerves.[10]

This event had two immediate consequences. In the first place Vesalius acceded to this 'request'; he raced into print in 1538 with not three but six great 'Tables', enormous woodcuts each printed as a single sheet. They were printed in the obvious place, in near-by Venice (from which Padua was controlled), which was one of the printing centres of the world. Vesalius wrote of the role he intended for them that

> I knew such drawings would be of no little help to those who had been present with the dissector. In other cases, indeed, I think that as it would be difficult so it would be pointless and quite impossible to want to acquire the knowledge either of the parts of the body or of the simples of the apothecaries from pictures alone or from formulae, but no-one would deny that drawings specially conduce to confirming the memory of things seen (*ad memoriam rerum confirmandam*). [I and the artist have done this] for the great advantage of those who judge it not only honourable and beautiful, but also useful and necessary to contemplate the skill and artistry of the Supreme Maker, and to look into that home of the soul (as Plato calls it).[11]

These Tables are on (and here I translate the titles):

1A. *The Liver*, the workshop of bloodmaking, which transfers the chyle from the stomach and intestines, and washes the melancholic humour out into the spleen. And
1B. The *Organs of Generation* of the male and female.
2. Description of the *Vena Cava*, by which the blood, the nutriment of all the parts, is spread through the whole body.
3. *The Great Artery*, arising from the left cavity of the heart, both bringing the vital spirit to the whole body, and tempering the natural heat through its contraction and dilation.
4, 5, and 6: *The Bones of the Human Body*, from the front, the side and the rear.

The skeleton was that of the first body he dissected at Padua. Vesalius credits one Joannes Stephanus of Calcar with drawing these skeleton pictures. Stephan was a Fleming, like Vesalius himself, and was in Venice as a pupil of Titian. (See Figures 4.3 to 4.8.) These tables were rapidly pirated by other printers. Sets are known to have been produced in the next few years at Paris, Cologne, Augsburg, Strasburg, London, Marburg and Frankfurt.[12]

Six Tables were issued, but Vesalius intended more. We know that at the beginning of January 1539 he had completed two more on *the nerves* (on the seven pairs of cranial nerves; on the small branches of the dorsal medulla). These he kept from print until he could finish yet further Tables, on the *muscles* and on all the *internal parts*. (The first of those on the nerves may be the source of the unique seventh Table pirated at Cologne, see Figure 4.9). All these he was intending to do with the help of the artist Joannes Stephanus, as soon as they had enough bodies for dissection.[13] So we know what Vesalius was busy at in early 1539: producing a whole series of Tables, based on the dissection of man and other animals.

The anatomy actually represented in the Tables that were printed has been authoritatively identified as a mixture of human, ape and other animal. Most striking from our point of view is the fact that the veins and arteries are dealt with as quite separate and distinct systems of vessels, respectively nourishing and vitalizing the body. In Table 2 the veins are shown as deriving from the liver, labelled by Vesalius as 'the principle [= beginning] of the veins'; the heart is not even illustrated on this Table! The arteries, by contrast, are shown in Table 3 as deriving from the heart, labelled by Vesalius as 'the flame of the vital faculty and the principle [= beginning] of the arteries'.[14] Thus the anatomy illustrated (and the actual variety in the animal sources of it) is in no way anti-Galen, but instead is just about as Galenic as one could ask for. The same is true of the legends printed down the side and keyed into the

TABULA I

PRAESTANTISSIMO CLARISSIMOQVE VIRO DOMINO
D. NARCISSO PARTHENOPEO, CAESARIAE MAIESTATIS MEDICO PRIMARIO.
Domino suo & patrono, Andreas VVesalius Bruxellensis S.D.

NON ita pridem, Narcisse doctissime, quum Patauii ad medicinæ chirurgicæ lectionem delectus, inflammationis curationem pertractarem, diui Hippocratis & Galeni de reuulsione ac deriuatione sententiam explicaturus, uenas obiter in charta delineaui, ita ratus quid per κατ' ἴξιν Hippocrates intellexisset facile posse demonstrari. Nosti namque quantum hac tempestate, ea dictio dissentionum atque contentionum, etiam inter eruditos, de uena secanda concitauerit, dum alii fibrarum consensum ac rectitudinem, alii aliud nescio quid, indicasse Hippocratem affirmant. Verum illa uenarum delineatio tantoperè medicinę professoribus studiosisque omnibus arrisit, ut arteriarum quoque & neruorum descriptionem, à me obnixè contenderent. Quia uerò ad meam pertinebat professionem Anatomes administratio, ipsis deesse non debui, potissimum quum scirem eiusmodi lineamenta, his qui secanti adfuissent, non mediocre commodum allatura. Aliàs siquidem aut partium corporis, aut simplicium pharmacorum cognitionem ex solis picturis, seu formulis uelle assequi, ut arduum, sic quoque uanum ac impossibile omnino arbitror: sed ad memoriam rerum confirmandam apprimè conducere, nemo negauerit. Cęterum cum plurimi hęc frustra imitari conarentur, rem prælo cõmisi, atque illis tabellis, alias adiunximus, quibus meum σκελετὸν nuper in studiosorum gratiam constructum, Ioannes Stephanus, insignis nostri seculi pictor, tribus partibus appositissimè expressit, magno sanè usu eorum, qui non modo honestum, aut pulchrum, sed etiam vtilè ac necessarium iudicant summi opificis solertiam artificiùmque contemplari, & domicilium illud animę (ut Plato ait) introspicere. Prętera singulis partibus, quãquam id in pręsenti negocio non admodum ex sententia confici potuit, sua nomina asscripsimus, barbaris, quæ etiam peritiores in plurimorum libris subinde remorari solent, minimè prætermissis. Quòd autem ad rei ueritatem attinet, nullum hic apicem ductum puta, quem Patauini studiosi in huius anni confectione, à me demonstratum non attestabuntur: ut interim sileam de Parisinis præceptoribus meis longe doctissimis & Louaniensibus medicis, apud quos non semel Anatomen publicè administraui. Porrò ut nouus hic noster conatus, alicuius patrocinio commendatior in lucem auspicato prodeat & ancipitem iudiciorum aleam securius experiatur, celebritati illum nominis tui nuncupare uisum est: partim quòd pręter incomparabilem uariarum linguarum cognitionem, eximiam quandam singularémque Anatomes, sicuti etiam medicinę & philosophię, scientiam adeptus sis: adeò ut meritò apud nationes omnes, tanquam pręcipuum peritissimorum medicorum & literatorum hominum, decus ac ornamentum ab eruditissimis quibúsque prędiceris: deinde quòd inter clarissimos uiros ea polleas animi prudentia, integritate, mira erga omnes naturę mansuetudine & gratia: ut CAROLVS QVINTVS Inuictissimus Romanorum Imperator semper Augustus, acerrimus ingeniorum æstimator, non suæ dumtaxat sanitatis tuendę, aut amissę recuperandæ præcipuum tibi locum concrediderit: uerum etiam te uniuersis Regni Hispanię ac Neopolitani medicis pharmacopolarumque officinis, in florente etiãnum ætate tua, ceu fidissimum censorem pręfecerit, compluribúsque honoribus & muneribus, inter tot præclaros alioquin uiros, amplissimè illustrauerit. Suscipe itaque Vir ornatissime hoc chartaceum munusculum, ea humanitate, qua me quondam excepisti, dum non exiguis beneuolentię signis, animum erga me tuum peculiariter declarasti: quod si gratum tibi ac studiosis fore intellexero, aliquando maiora adiiciam. Vale Patauii Calē. Apri. An. salutis. M.D.XXXVIII.

IECVR SANGVIFICATIONIS OFFICINA, PER VENAM PORTAM, QVAE GRAECIS φλὲψ πύλη, Arabibus vero [illegible] varidhescoer appellatur, ex ventriculo & intestinis chylum trans sumit, ac in lienem melancholicum succum expurgat.

A Cauum, seu simum iecoris.
B Vena porta, iecoris manus.
C Ramuli in flauæ bilis vesiculã.
D Ad pancreas & ecphysim, seu duodenum intestinum.
E Ad dextrum gibbi ventriculi.
F Ad dextrum fundi ventriculi & superiorem omenti membranam.
G Portę bifurcatio maxima.
H Per omenti inferiorem membranam & pancreas delata, varie diffunditur.
I In omenti membranam inferiorem, parte dextra.
K Per ventriculi cauum, eius os tandem numerosis propaginibus amplectens.
L In membranam omenti inferiorem parte media, quę primum in duas, deinde in plurimas exiguas venulas diuaricatur.
M Multifariam diuisa, per rectam lineam lienis simo implantatur: hac fęculentus sanguis in lienem transmittitur.
N Vtraque ad ventriculi gibbi sinistrum, & secunda satis obscurè ad ventriculi os procedit.
O In sinistrum fundi ventriculi, & superiorem omenti membranam: hac non mediocrem excrementi lienis portionem in ventriculum excerni putauerim.
P Numerose inter mesenteri membranas distributa in intestina excurrit: ab hac ne, an à caua, hęmorrhoides sint: non ausim certo affirmare. Nam ex vtraq; vena rami in eam partē ptinent, & etiã maiores a porta: nec per portam melancholicum sanguinem expurgari, fortè alienum animal vertentis, apparebit.

GENERATIONIS ORGANA, SVPERIVS VIRI, INFERIVS MVLIERIS.
Tertia figura semen deferentium vasorum implantationem refert.

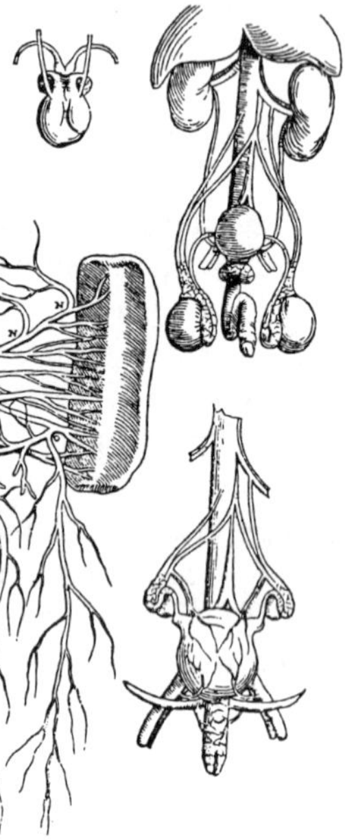

GALENVS VENAE PORTAE RAMOS PRAECIPVOS SEPTEM ENVMERAT.

I.

4.3 Table one of the *Tabulae Sex*: on 'The liver, the workshop of bloodmaking …', and 'The organs of generation'.

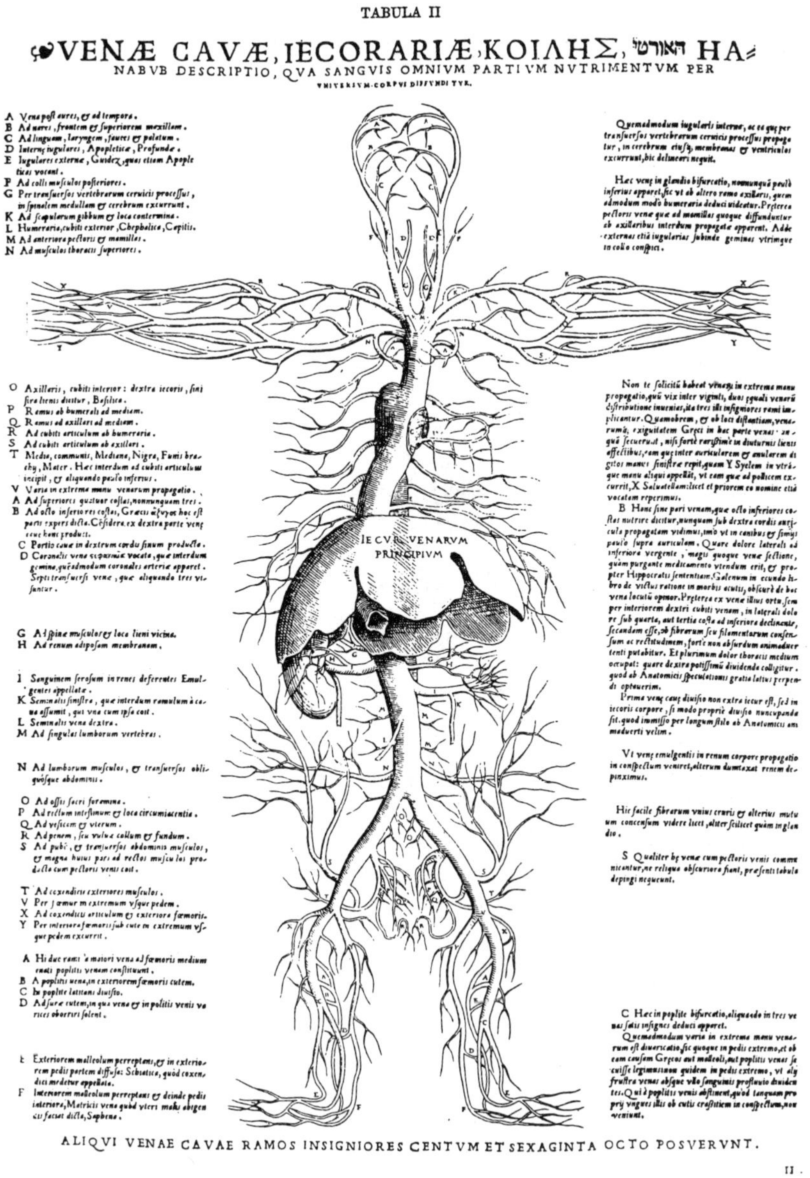

4.4 Table two of the *Tabulae Sex*: 'Description of the *Vena Cava*, by which the blood, the nutriment of all the parts, is spread through the whole body'.

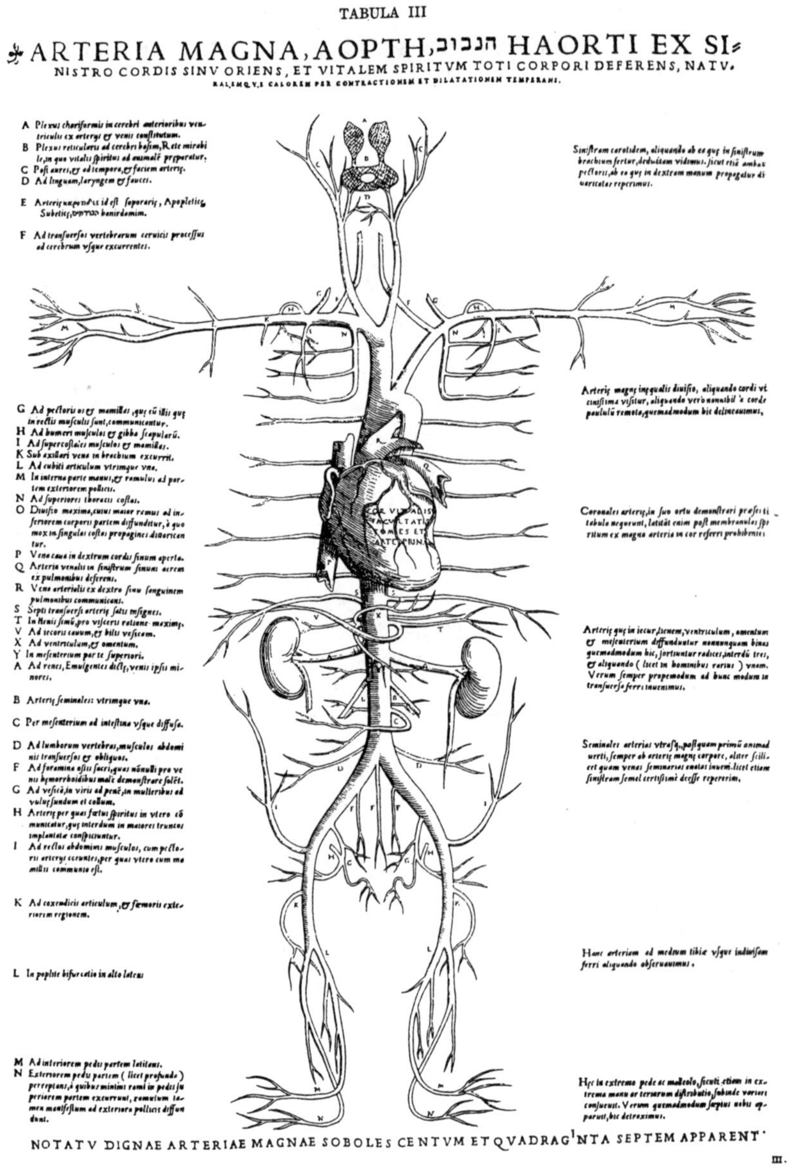

4.5 Table three of the *Tabulae Sex*: 'The *Great Artery*, arising from the left cavity of the heart, both bringing the vital spirit to the whole body, and tempering the natural heat through its contraction and dilation'.

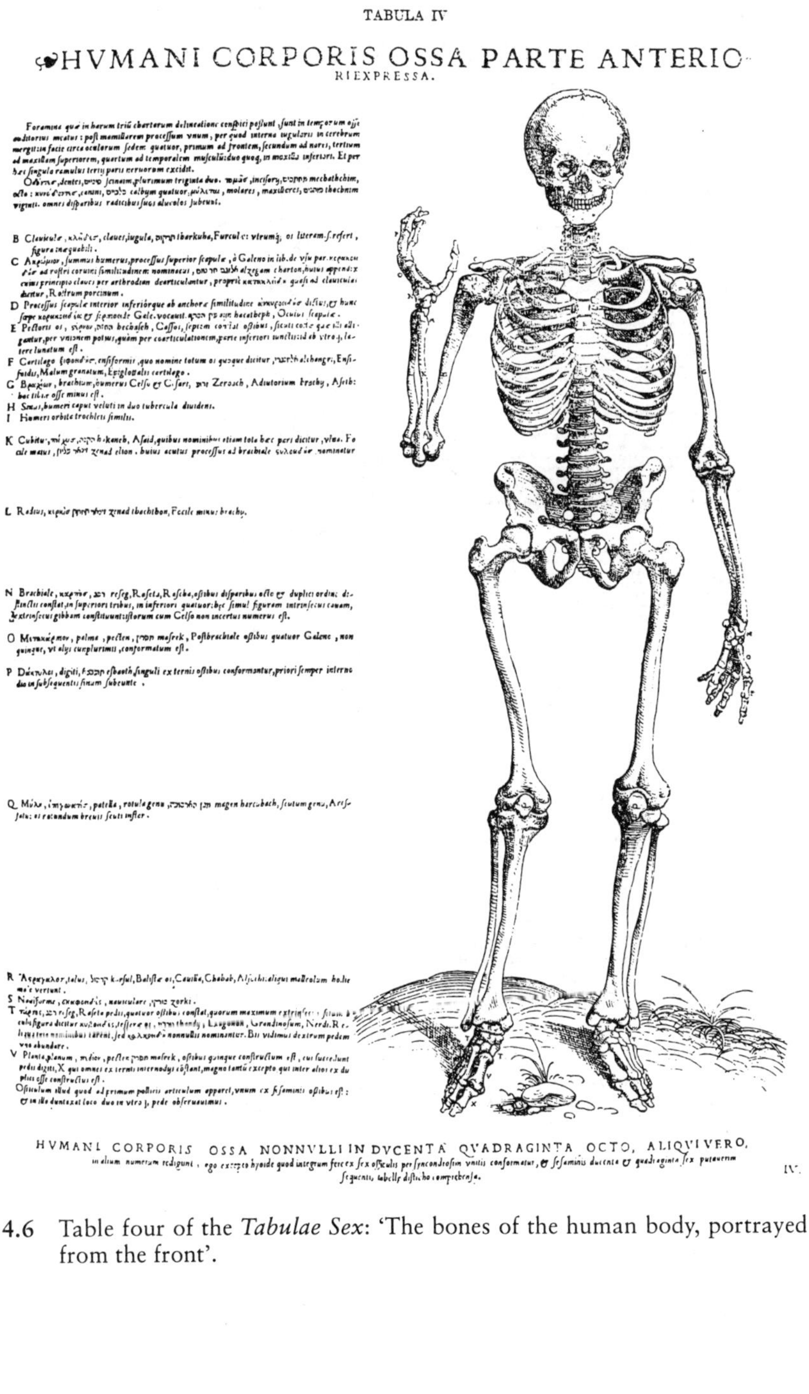

4.6 Table four of the *Tabulae Sex*: 'The bones of the human body, portrayed from the front'.

TABULA V

LATERALIS ΣΚΕΛΕΤΟΥ FIGVRÆ DESIGNATIO.

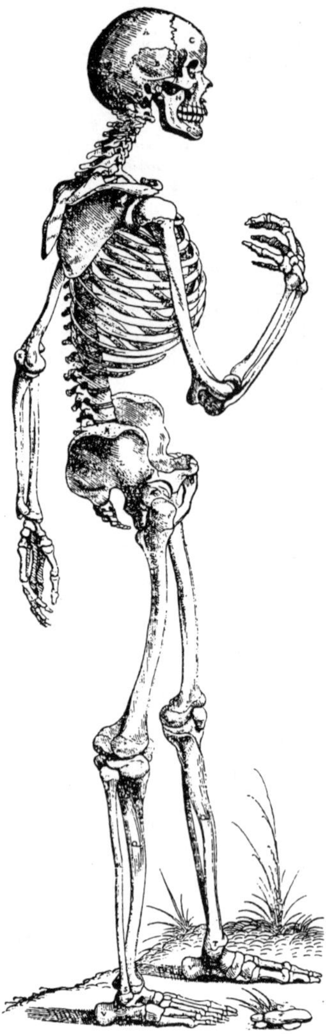

DISTICHON OSSIVM NVMERVM COMPLECTENS.

Adde quater denis bis centum senáque, habebis
Quàm sis multiplici conditus osse semel.

V.

4.7 Table five of the *Tabulae Sex*: side view of the skeleton.

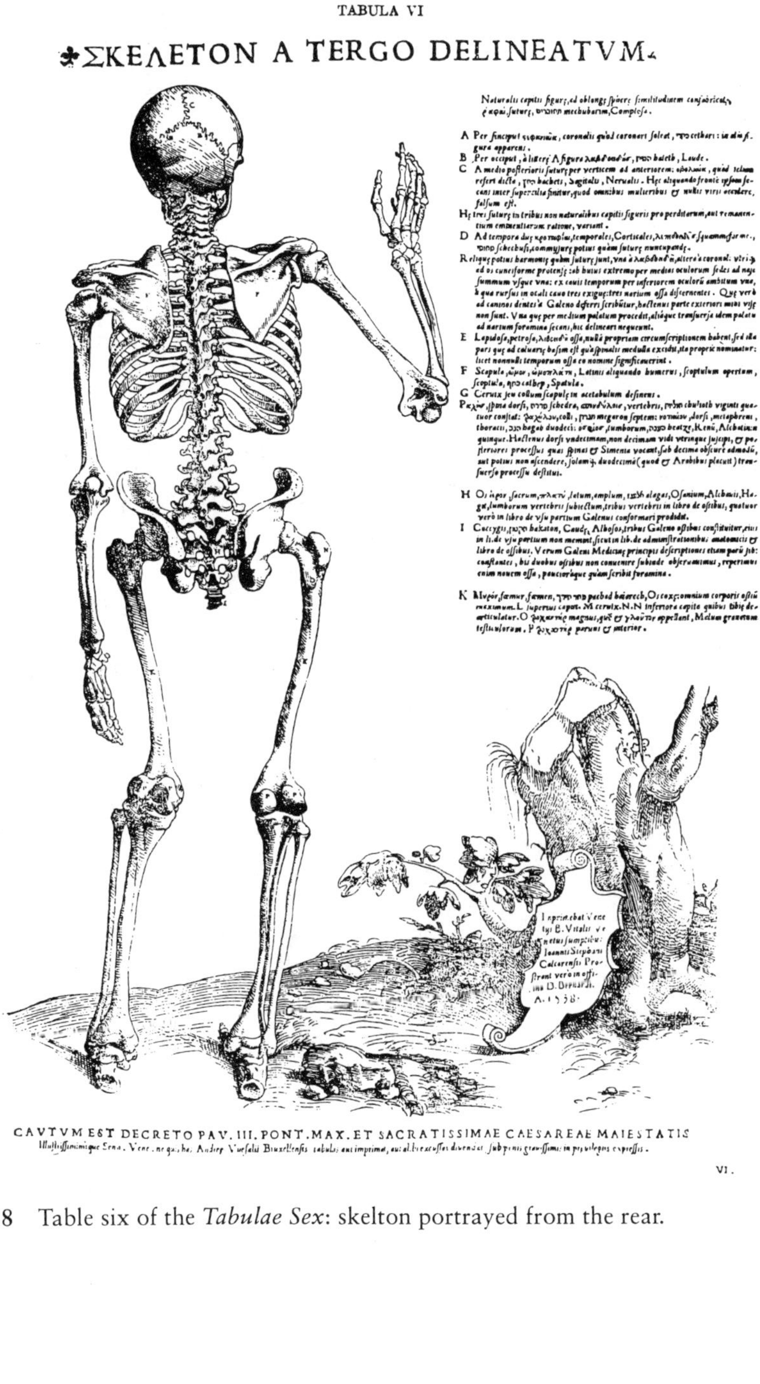

4.8 Table six of the *Tabulae Sex*: skelton portrayed from the rear.

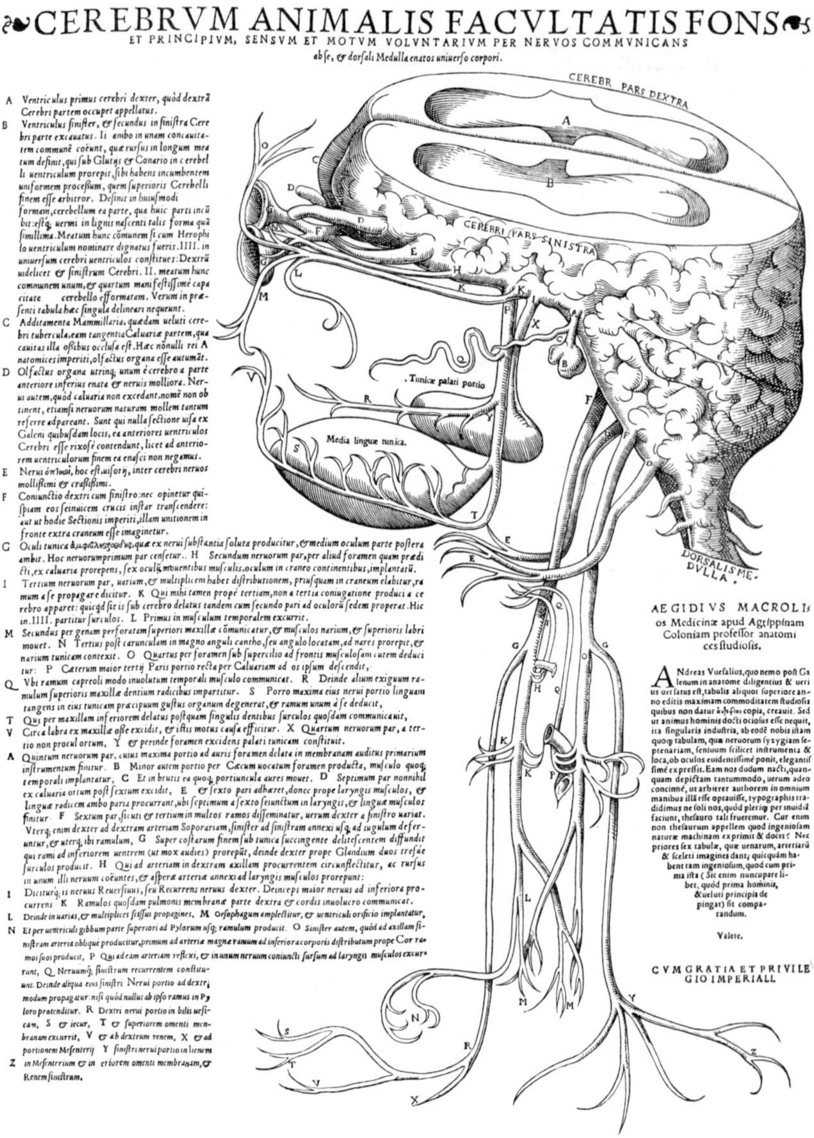

4.9 Unique copy of a seventh table, 'On the brain as the source and origin of the animal faculty, communicating sense and voluntary motion through the nerves'. Cologne, 1539.

parts. If we take Table 3 for instance, there happily represented is a structure not actually present at the base of the brain in man, but identified by Galen: the *rete mirabile*. And the central legends reveal a perfectly Galenic view of the roles of the vessels of the heart. They read:

> *P*: The vena cava, open into the right cavity of the heart;
> *Q*: The 'artery-which-looks-like-a-vein', bringing the air from the lungs to the left cavity;
> *R*: The 'vein-which-looks-like-an-artery', imparting blood from the right cavity to the lungs.

These Tables are novel in two ways. First, that they illustrate Galenic 'systems' (as we would now call them): the veins, the arteries, the nerves, and also the bones; and this is quite different from the 'three-venter' approach. We shall return to this innovation. Second, that they were intended as teaching aids, literally as aide-memoires. But it was not illustration as such, not even illustration on such a grand scale, which was Vesalius's great Renaissance innovation, striking as these pictures were to his contemporaries and still are to us.

The other consequence of that initial production of a picture of the veins at his lecture, was that Vesalius became publicly embroiled in a hot dispute about where and how it was best to bleed, as a means of cure, someone who had pleurisy. By the beginning of 1539 he had himself entered the fray, with a small but offensive book on the subject.[15] The rights and wrongs of this dispute need not detain us, except to note that it was rightly treated as a serious issue, with people's lives depending on the answer. As a dispute it had begun some 25 years before, as the Greek text of Hippocrates and the Greek text of Galen's exposition of Hippocrates had become available; battle was joined between those who supported the traditional (the 'Arab') way, and the Moderns, those who preferred the new (the 'Greek') way. But then discord struck in the camp of the Moderns. Everything seemed to hang on what Hippocrates meant by the term *kat' ixin*, 'in a straight line'. It was at this point that young Vesalius entered the lists, as a Modern. He claimed that if this term meant what it ought to mean, then Galen's exposition of it was wrong. Either Hippocrates or Galen must be at fault, and it was unthinkable that Hippocrates was wrong. Now this was quite an astonishing thought: that Galen, the Prince of Physicians, might be wrong, for the generation of Vesalius's teachers (as we have seen) had dedicated themselves to following Galen as closely as they were able, as source of truth. But Vesalius, if we trust his own account, was not trying to castigate Galen. Instead he was doing a very Galenic thing: he was working from the evidence of the body – from the incidence and distribution of the veins, as he found them in the dissection of human bodies. In his eyes it all came down to a matter of which vein or veins fed the thoracic ribs: the elucidation of the doctrines of Hippocrates and Galen on this matter must start from here. And what Vesalius saw in the body was a single vein, the 'vein without pair', feeding all the thoracic ribs. This vein is letter B on Table 2 (Figure 4.4): just above, and just opposite where the vena cava is open to the (absent) heart. He drew his conclusions about bleeding in pleurisy accordingly; as he put it in the title of his book: 'that the axillary vein of the right elbow is to be lanced in pain in the side (*dolor lateralis*), and the melancholic humour is to be purged from those branches of the vena cava tending towards the seat of the disease'.

Now this placing of the 'evidence of the body' over the evidence of texts, was a most natural position for Vesalius to take. For what Vesalius excelled all his contemporaries at was dissecting: the evidence of dissection therefore was the paramount evidence to him, at least when he was weighing up the arguments on either side of a live dispute, for such a dispute could be resolved by the senses, by looking and by feeling. And it was not just the evidence of Vesalius's own senses which could be appealed to, but the evidence of the senses of all those trustworthy witnesses who had been present at his dissections at Paris, Louvain or Padua: 'No-one who has either engaged in the task of dissecting personally or who has observed a dissection will allege that this vein arises under the base of the heart or before the vena cava reaches the heart.'[16] If he could get everyone else to concede that the evidence of dissection was the arbiter of the dispute, then Vesalius, the 25 year-old dissector, had won the argument over all the eminent professors and practitioners. This is why he had drawn the picture of the veins in the first place.

(ii)

It is at a series of demonstrations he gave in Bologna in January 1540 that we can best see Vesalius in action before he had his 'blinding flash', as it were. He was invited to Bologna by the students, who still retained the right of organizing the dissection.[17] His reputation had preceded him, but it was because of his reputation as an exceptionally skilled dissector (with an unusual claim about bleeding in pleurisy), not because he was some kind of reformer of anatomy that he was invited. We are most fortunate that a student (a German) took down virtually verbatim notes of this event, including comments on the reaction of the audience. This allows us to see with extraordinary clarity the way in which a public anatomy was conducted in 1540, and precisely what role the dissector was expected to fill.[18] Thus we can see that the event as a whole consisted of:

1. A series of *lectures*, based on the *Anatomia* of Mundinus, given on this occasion by the Professor of Theoretical Medicine, and
2. In a separate place, and at a separate time, an interlocked series of *demonstrations* of the parts dealt with in the immediately previous lecture, conducted on a series of human and animal bodies.[19]

The lectures were given by Professor Matthaeus Curtius (Matteo Corte, *c.* 1475–*c.* 1542), a grand old man of academic medicine.[20] Curtius had received his own education when the scholastic approach was still the norm. However, when obliged by custom and statute to lecture on the text of Mundinus, Curtius showed himself in anatomy to be just about as up to date in his attitudes as could have been wished. In lecture after lecture he announces his position: that wherever Mundinus had deviated from the strict doctrines of Galen, Mundinus was in error. That is to say, for Curtius the standard of truth was the doctrine in the pristine text of Galen. In this he was as modern as Sylvius or Guenther. However, today Curtius might look to us like an anatomical reactionary and ignoramus, for it was his unlucky fate to have had Vesalius as his demonstrator on this occasion. It was unlucky in two ways. First, Curtius was a participant (as a Modern) in the debate on where to bleed in pleurisy, and had published on it in 1532. Thus he and Vesalius were at daggers drawn on a highly controversial issue. Second, Vesalius was busy transgressing all the boundaries of what a demonstrator should do. He was thus challenging Curtius on an issue of etiquette, respect and authority. And in so doing the 25-year-old Vesalius was to subject the 65-year-old Curtius to the most outrageous public humiliation. The students, of course, loved every minute of it.

Curtius had already given five lectures (covering Mundinus's text up to the peritoneum, the membrane enclosing the organs of the lower venter) before the first of the demonstrations. Meanwhile (though it is not recorded in this student's notes) Vesalius was lecturing on Galen's book *On the bones*.[21] It was only during the course of the third lecture that the first criminal who was to be dissected was executed.

Picture the scene. We are in the middle of the fifth lecture in the church of San Salvatore. The beadle arrives with a message from the students in charge of the demonstration: the body was now prepared for the anatomy.

> And they asked us through the beadle, that after the lecture we should go to the anatomy in good order and without disturbance: for we should all well be able to see wonders which we had not seen hitherto. Dominus Curtius likewise asked us to do so. And yet after the lecture we proceeded to the demonstration in great disorder, as the mad Italians do.[22]

The place they rush to, where the body is laid out, is another church, San Francesco. Vesalius has had four banks of temporary staging erected in a circle to create within the church a theatre to accommodate about 150–200 people.[23] It is the kind of theatre of which Benedictus had written. The students running the demonstration go in first; then those who had paid 20 sol; then the doctors and the students. At last Vesalius arrives.

> Then Dominus Vesalius began: Domini, he said, you know how the Physicians, both ancient and modern, usually divided the human body: the Egyptians and the Arabs begin from the trunk and the extremities; but Galen, whom Mundinus has followed too, divides it according to the three venters . . .[24]

Vesalius is obviously beginning to give a lecture of his own. Immediately Curtius interrupts, and requests Vesalius to proceed with demonstrating what he, Curtius, has been lecturing on. Vesalius recovers his ground: 'But, omitting these issues, we shall proceed to our anatomy.' And he does so. In the lower venter he demonstrates the skin, blistering off the outer layer to reveal the inner; then the fat under the skin; then the muscles of the belly, which he has dissected out in readiness. But he cannot resist talking and giving his opinion:

> he showed us two long muscles along the length of the belly, joined together at the lower part by white membranes, but they were more separated at the top end. And he showed us how they ended at the sternum (*os pectinis*), but had their origin from the collarbone (which could not yet be seen) and not from the shield of the mouth of the stomach as Mundinus says.[25] (See Figure 4.10)

Curtius interrupts again: But this is not the opinion of Galen! Vesalius answers:

> No, Dominus, he said, and even if Galen says that, yet we shall demonstrate here that in fact it is so. But, he said, we do not now want to fight with many words.
>
> Then the [student] Rector said, rather uncouthly: Dominus Doctor Andreas, do not be afraid of telling your opinion on these matters, do not fear such Fathers [as Galen] (*patres istos*).
>
> Vesalius answered: Later in the anatomy of the middle venter I shall demonstrate that these long muscles begin from os iuguli and not from the mouth of the stomach.[26]

As we can see from the first moments of the demonstration, one thing Vesalius was not afraid of was giving his own opinion – whether it was at variance with that of the Professor, or with that of Galen. But Vesalius, as the demonstrator, was not supposed to be giving any opinion at all; it was the job of the lecturer to lecture, not that of the demonstrator. But there was to be no stopping Vesalius. Vesalius was

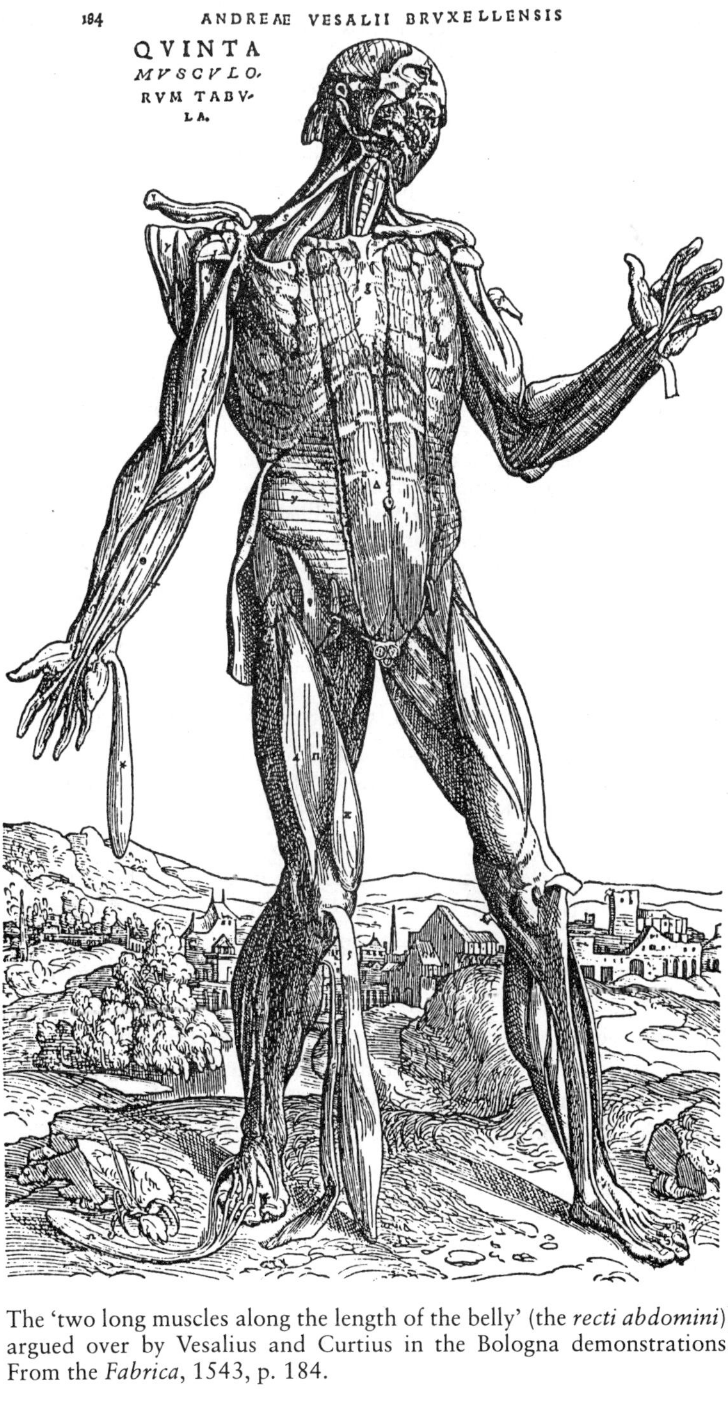

4.10 The 'two long muscles along the length of the belly' (the *recti abdomini*), argued over by Vesalius and Curtius in the Bologna demonstrations. From the *Fabrica*, 1543, p. 184.

beginning to reveal himself as possessed of a talent, an ego, and a lack of false modesty comparable only to the great Galen himself.

Right from this first demonstration we can see how Vesalius conceived his task as demonstrator, what he thought ought to be demonstrated and in what order. He had near-ideal conditions here in Bologna, with three human bodies and several dogs and other animals available to him over the course of this series, lasting two weeks. He had already demonstrated the *skeleton.* Now, in and from this first demonstration on the body what Vesalius is obsessed by is the *muscles*.

> Observe, he said, that all muscles take their rise from bones, beginning from sinews (*a nervis*) where their head is, and then ending in tendons or cords they are again fastened to bones, in order to effectuate their natural voluntary movement in the body . . .
>
> They had killed a dog upon which he showed that the muscles in dogs as also in other animals were fastened in quite another way [than in man], so that they should be able to run faster. He promised to show us after dinner the anatomy of the muscles of one forearm; in the meantime we ought to read about them in Galen, *De usu partium* I and II, and *De anatomicis administrationibus* I. He had erected his anatomy of the bones [i.e. a skeleton] on the table, by means of which he always the better demonstrated to us the site of the parts of the body.[27]

Vesalius has brought an articulated skeleton with him, and it is present at all the demonstrations (see Figure 4.11). He seems to have travelled nowhere without at least one skeleton for demonstration purposes (he had actually brought a spare one, plus the skeleton of a monkey, with him to Bologna, both of which he gave to a friend). Every time he discusses a muscle or group of muscles, he shows where and how it is connected to the bones. He has taught the *bones* from the skeleton. And he dedicates the first corpse to a full demonstration of the *muscles*.

Thus what happens for the first ten days is that Curtius goes on correcting Mundinus morning and evening in his lectures, as he works through the account of the three venters. When the students come to the demonstration Vesalius gives a short treatment, but not precisely of the parts that Curtius has dealt with. Where possible Vesalius does his demonstration on the human corpse, but where the pertinent parts had already rotted (such as the intestines), he uses a dog or some other animal. But the bulk of each demonstration is taken up by the dissection of the *muscles* – that is, things which Curtius had not been dealing with. First the muscles of the inner forearm:

> He showed by dissection how these muscles arise from the head of the bones of the forearm with a thin sinew (*nervo*), and how after a long course they end in the hand and the joints of the fingers, how

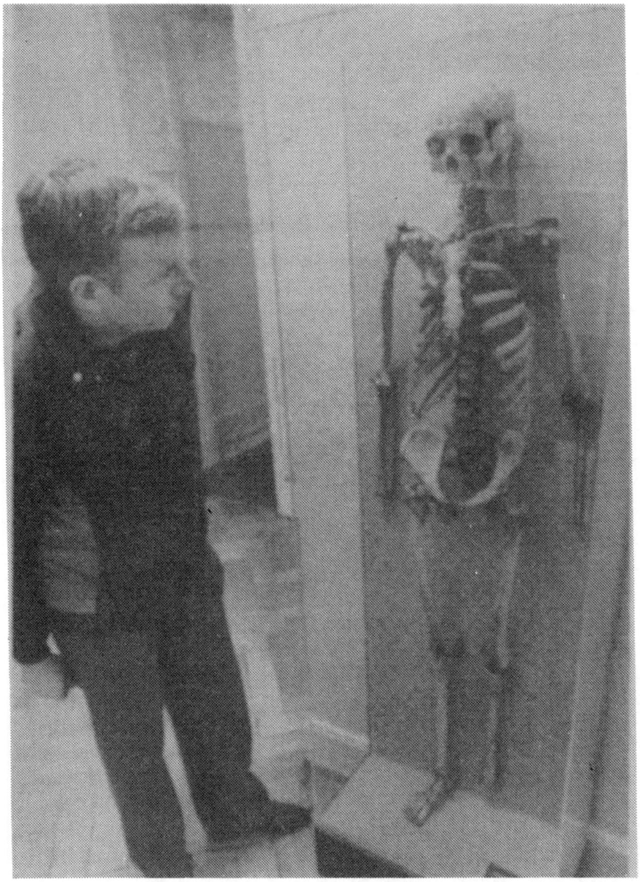

4.11 The skeleton that Vesalius prepared at Basel, now the oldest anatomical preparation in existence.

> the muscles are situated one over the other in a double arrangement, always four over four, and how the lower ones extend to the first joints, while the upper ones extend to the second and third joints, always passing through the first muscles. This was indeed most beautiful to see.[28]

This, as we shall see, was a very significant place to start. The dissection of these muscles Vesalius was to take as his trademark (see Figures 4.12 and 4.13). Then (demonstration 3) he dissected the muscles of the outer forearm; of the palm of the hand (4); of the face, forehead, jaw, eyes, nose, lips (5); of the shoulders, shoulder-blades, neck and upper arm (6); of the rest of the arm (7); of the thorax (8); of the hip and upper leg

ANDREÆ VESALII.

4.12 Vesalius showing the dissection of the muscles (tendons) of the human forearm. From the *Fabrica*, 1543, preliminary matter.

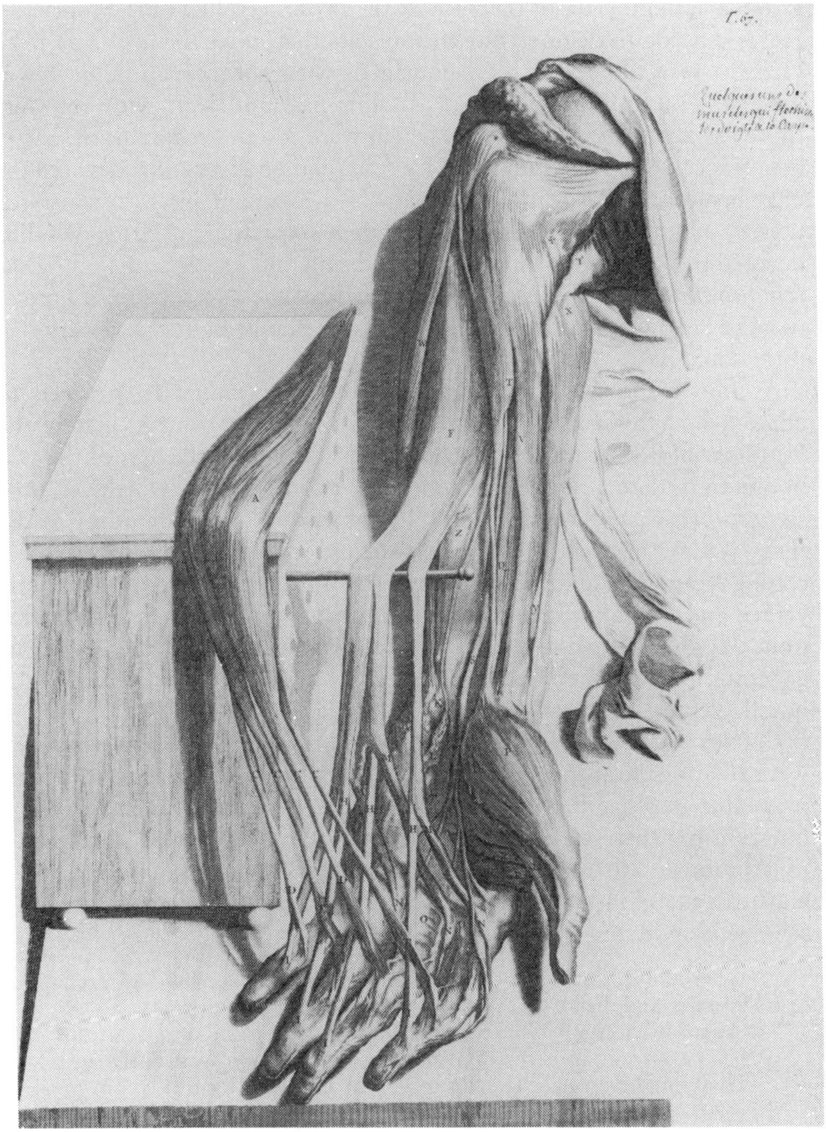

4.13 The flexor muscles of the lower arm, separated from each other. This reveals in greater detail the muscles that Vesalius is shown dissecting. Drawing by G. de Lairesse, for G. Bidloo's *Anatomia Humani Corporis*, Amsterdam 1685. Printed in William Cowper, *Anatomia Corporum Humanorum Centum et Quatuordecim Tabulis ... Explicata ...*, Leiden, 1739.

(9); of the thigh (10); of the back of the lower leg (11); of the foot (12). He used a dog to demonstrate the muscles that move the neck and head (13) because these had been damaged when the human subject was hanged. The only muscles that Vesalius had not dealt with yet were those associated with the penis, which were too desiccated in the available subject. While waiting for a fresh human body Vesalius devoted the next demonstration to the dissection of a pregnant bitch to show the uterus, and of her unborn pups. With a fresh human body, Vesalius devoted the 15th demonstration to the brain and its parts, including the *rete mirabile* from the brain of a sheep, and the seven pairs of cranial nerves. This demonstration led to the students behaving in a very disturbed manner.

At this point, Vesalius announced that he wanted to proceed to anatomizing the *arteries*, the *veins* and the *nerves*. Unfortunately the Professor of Practical Medicine had sneaked off with the best of the two bodies to initiate a rival dissection. So as this body was 'worthless' (*hoc subjectum nil valet*) Vesalius took it and started on the anatomy of the *three venters* instead. Meanwhile Curtius, steadily following and correcting Mundinus in his lectures, was still at the beginning of the middle venter and had almost reached the lung. Vesalius devoted demonstrations 16 and 17 to the organs of the lower venter and the organs of generation (the subject was male); 18, 19 and 20 to the organs of the middle venter.

Curtius and Vesalius were now, for the first time, in step in what they were treating. And they had now reached the point where conflict was inevitable in what they were teaching. For here they had reached the question of where to let blood in pleurisy. Vesalius gave a demonstration (number 20) to familiarize the students with his point of view; Curtius in response devoted two lectures to a refutation of it, in full scholastic style. But then:

> When the lecture of Curtius was finished, Vesalius, who had been present and heard the refutation of his own arguments, asked Curtius to accompany him to the anatomy [the 22nd of the series]. For he wanted to show him that his own opinion was most true. Therefore he brought Curtius to our two subjects for dissection.
>
> Now, he said, excellentissime Domine, here we have our subjects: we shall see whether I have erred. These we now want to examine; and we want to put Galen aside for a while, for I acknowledge that I have said – if it is permissible to say so – that here Galen has erred because he did not know the site of this 'vein without pair' in the human body, something we have today the same as he did.
>
> Curtius answered by smiling, for Vesalius, as he was choleric, was very excited: No Domine, he said, Galen should not be put aside by us, since he always well understood everything, and conse-

quently we also follow him. Do you know how to interpret Hippocrates better than Galen knew?

Vesalius answered: I am not saying that, but I am showing you here in these two subjects the 'vein without pair', how it nourishes all the lower ribs, except the two upper ones in which pleurisy does not occur. For always here – here he knocked with his hands against the middle of the chest – occurs inflammation and pleurisy, not in the two upper ribs. Consequently, as this vein is also distant from the heart, as you see, by three fingers' breadth, it will in pleurisy and all pain in the side always be better to bleed from this vein only; or it ought to make no difference from which part [of the body] blood is let, because the ribs are nourished exclusively by this vein.

Curtius replied: I am no *anatomista*, but there can well be still other veins nourishing the ribs and the muscles besides these.

Where, I ask? Vesalius said, Show them to me.

Curtius said: Do you want to deny the ducts of Nature (*meatus naturae*)?

Oh! Vesalius said, you want to talk about the invisible and the occult – but I am talking about the visible (*de manifestis*).

Curtius answered: Indeed I always deal with the most evident (*de apertissimis*). Domine, you do not understand Hippocrates and Galen well on these matters.

Vesalius replied: It is quite true, because I am not such an old man as you are, etc.

Thus with much quarrel and scoffing they attacked each other, and in the meantime they accomplished nothing.

Vesalius said: Domine Doctor, I beg Your Excellency not to think me so unskilled that I do not know and understand these matters.

Smiling, Curtius said: Domine I did not say so, for I have always said that you are excellent, but I have rejected your false exposition of Hippocrates implying that Galen should have erred in these matters.

Vesalius replied: I acknowledge that I have said that Galen has erred in these matters, and this is evident here in these subjects, as also many more mistakes of his.[29]

Aliud est monstrare, aliud vero est inferre, it is one thing to show, but it is quite another to draw conclusions.[30] This is what they are arguing about, the difference between 'what is visible' (*manifestus*) to the eyes and 'what is most evident' (*apertissimus*) to reason as taught by authority.

Throughout these demonstrations one can see a number of distinctive features of Vesalius's approach. Firstly, as with his first known lecture at Padua, he thinks visually all the time. He is forever picking up a piece of charcoal and drawing on the dissection table; our student reporter remarks on Vesalius doing so on several occasions, copying down the spontaneous sketches into his own notebook. Vesalius also brings pictures into the dissecting theatre: his own Tables, his little book on

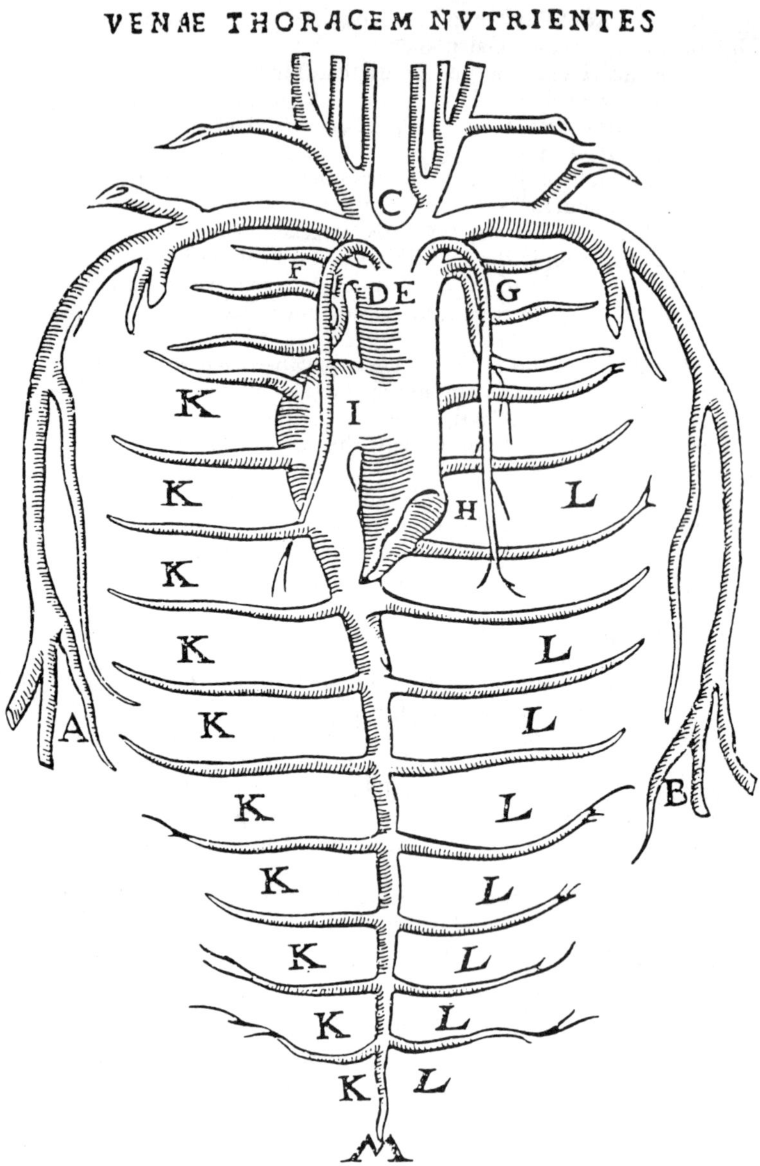

4.14 The veins nourishing the thorax. From Vesalius's book on bloodletting of 1539.

bloodletting (which has a picture of the ramifications of the 'vein without pair' in it, see Figure 4.14), and a book recently published in Marburg by Dryander,[31] with illustrations of the dissection of the brain.

So whether or not Vesalius was the artist for his own woodcuts, we can see that he instinctively thought in terms of pictures, diagrams and sketches. He had a pronounced visual sense, words alone were not sufficient for him.

Secondly, in these demonstrations we can see Vesalius having, or progressively acquiring, the courage (or impertinence) to criticize Galen relatively freely. As we saw when he was arguing with Curtius, Vesalius could now claim that comparison with the body revealed '*many more* mistakes' on Galen's part (*plures sui errores*). Yet Vesalius's criticisms are in fact very limited and specific. What he criticizes are *particular* doctrines, claims, statements by Galen (or in the texts ascribed to Galen).

What we have to recognize here is that Vesalius's experience of comparing Galen's texts with the body – the human body – that those texts were about, was second to none. Of the texts Vesalius is now able to compare to the body, some of them, which had been almost completely unknown to the generation of his teachers, he had got to know from editing them for the press. For the great *Complete Works* of Galen put out by the printer Guinta in Venice in 1541,[32] Vesalius had edited (sometime over the period 1539–40) *On the dissection of the nerves* and *On the dissection of the veins and arteries*. He had also revised Guenther's translation of Galen's *Anatomical procedures* for the same edition. But this work, in Vesalius's case, was not something which could be done simply in a study or a library. Vesalius could not keep away from the practice of dissection; everything he read about in Galen's anatomical works he wanted to dissect out and see for himself in the body. Thus when Galen writes, as he does at the beginning of *Anatomical procedures*, that one should study human bones at every opportunity and preferably the articulated skeleton, this is precisely what Vesalius does. When Galen writes, in the same place, that one should study every muscle with relation to its origin and insertion (that is, how it is connected to the bones), this is what Vesalius does.

The sequence of dissecting that Vesalius adopted in the Bologna demonstrations – and which conforms to his repeated exhortations about the right sequence to pursue – is equally striking. It is:

1. The Bones (using a skeleton)
2. The Muscles; with the tendons, ligaments, etc
3. The Arteries
4. The Veins
5. The Nerves

6\. 7. 8. The organs of the lower venter; of the middle venter; and of the highest venter.

We have already seen something of this schema in the *Six Tables*. The unexpected disappearance of one of the bodies thwarted Vesalius in carrying this out at Bologna, and he had to omit the full separate treatment of the arteries, of the veins and of the nerves. But why did he choose this sequence? There was nothing in the statutes about it, and it was not in line with the 'Mundinus' sequence. Quite simply, because this is precisely the dissecting sequence that Galen recommends, as we saw above in Chapter 1.

Before we leave these marvellous demonstrations, it is important for us to look at the last of them, the 26th. First Vesalius shows the anatomy of the eye, using that of a sheep, cutting it open and shaking out into his hand the three 'humours', the white of the eye (the sclera), the vitreous and the crystalline. 'All these,' he said, 'anybody can see for himself at home; for surely, Domini, he said, you can learn only a little from a mere demonstration if you yourselves have not taken the objects (*subjecta*) into your own hands.'[33] Our notetaker frequently followed this advice; at the end of demonstrations he had gone up to touch and hold the penis, the lung, the veins of the arm, the *rete mirabile* and dissected head of various of the corpses.[34]

But Vesalius has left the *coup de theatre* until last.

> Finally he took a dog ... He bound it with ropes to a small beam so that it could not move; similarly he tied its jaws so that it could not bite. [See Figure 4.15] Here, Domini, he said, you will see in this living dog what the function (*usus*) of the nervi reversi is: you will hear how the dog will bark as long as these nerves are not injured. Then, I shall cut one nerve, and half of the voice will disappear; then I shall cut the other nerve, and the voice of the dog will no longer be heard. And he did so; when he had opened the dog he soon found the nervi reversi around the arteries, and all happened as he said. The bark of the dog disappeared when he had by turn cut the nervi reversi, and only the breathing remained.[35]

The nervi reversi (the recurrent laryngeal nerves) are illustrated in the nerve Table (see Figure 4.9); they had acquired their name from the fact that they reverse their direction, first descending and then re-ascending.

> Finally, he said, I shall proceed to the heart, so that you may see its movement, and feel how great its warmth is, and thirdly so that you may feel here, around the ilium, the pulse of the artery with one hand, and the movement of the heart with the other, and you may tell me what its movement is: whether the arteries are compressed when the heart is dilated? or whether they have the same movement as the heart at the same moment? I myself saw how the heart of the dog bounded upwards; and when it no longer moved, the dog instantly died. Those mad Italians pulled the dog back and forth so that nobody could truly feel those two movements. But some students asked Vesalius what the truth about these move-

4.15 How to prepare an animal for vivisection. From the *Fabrica*, 1543, p. 661.

> ments is, what he himself thought, whether the arteries follow the movement of the heart, or whether they have a movement different from that of the heart. Vesalius answered: I do not want to give my opinion, you yourselves should feel with your own hands, and trust them.[36]

This is all very exciting; it is brilliant theatre. In its way it is also very moving: feel with your own hands, and trust them. But why was Vesalius performing vivisection demonstrations at all? Why these particular ones? And why was he saying 'feel with your own hands, and trust them'? Because Vesalius as vivisectionist was simply Galen restored to life. The use of vivisection to demonstrate certain phenomena was a technique employed to stunning effect (if we may trust his own account of it) by Galen in second-century Rome. And as we saw when discussing Galen, these particular things – the recurrent laryngeal nerves, whether the arteries pulsate of themselves or from the impulse of blood into them – these were Galen specialities, to make Galen points against Erasistratus. And the rhetoric of participation, feel with your own hands and trust them, this too is the Prince of Physicians himself speaking. All this had been pioneered by Galen and performed by him publicly in Rome fourteen hundred years before.

The Vesalius we have discovered so far is a northern youth, a resourceful Fleming, one who had become celebrated as a dissector in Italy, indeed at the most famous school of medicine in the world. He is someone who, by personality, is obviously prepared to criticize his elders (such as Curtius) in public, and his greater elder too (Galen). He stresses to his students the need for *personal engagement* in anatomy, and says that such personal experience is essential for anyone who claims to be a practitioner of medicine, and he puts the evidence of sight and touch above that of logic, custom and authority. All of this suggests that he is precocious, aggressive and intelligent. Yet all the striking and seemingly innovative things we have so far seen in Vesalius's approach to anatomy can, as we have begun to see, be laid directly at the door of Galen himself. For with all his innovations Vesalius was most certainly criticizing neither Galen the anatomist nor the project that Galen had been engaged in: Vesalius was not saying (for instance) that we need to find out about anatomy in some way, or in some terms *other* than those Galen used. Far from it. For the anatomical project of Galen is precisely what Vesalius was following. What he was criticizing was not Galen, nor Galen's project, but the points at which Galen himself had not fulfilled it properly. No one since Galen himself had followed the *practice* of Galen in anatomy as precisely as Vesalius was now doing. And in his grace, skill and expertise in dissecting, in his familiarity with it, in his making the senses the arbiter of truth, in his demanding to see *in the body* the pathways that other people invoked in their explanations, Vesalius was also another Galen. In his confidence in his own skills, his assertiveness, his willingness to criticize other people, Vesalius was another Galen too. From this we can see that Vesalius wanted to be, and felt himself to be, a second Galen. He was the first person since Galen's time to try to live out the experience of *being* Galen the anatomist. He was not just following what Galen said one should do in anatomy, but trying to *be* Galen in the present. *This* was the new Renaissance achievement of Vesalius: to *emulate* Galen the Ancient, in his own *practice* – not just to acquire a firm grasp of Galen's doctrines by consulting the dissected body with book in hand, as Sylvius and Guenther had sought.

This was Vesalius's ambition, and he was already putting it into practice in 1540. But the logic of this goal was to take him one further step, as yet unsuspected by him. He was to move on from 'being' Galen and doing what Galen did, to 'being' the Galen that Galen had intended to be, but had been unable to be. In this Vesalius was to go beyond Galen.

(iii)

What was that further step, and how and why did Vesalius come to take it? Since he was first a student and became interested in anatomy, Vesalius had been involved, like his teachers, in recovering to sight the body that Galen had described. But, I have been arguing above, Vesalius – unlike his teachers – was obsessed with doing that in a way as similar to that of Galen as possible. So Vesalius wanted to see the body as Galen had described it: the same body. But there was something radically different about the bodies that Galen and Vesalius as anatomists each had in front of their eyes. Vesalius's further step consisted of being able to notice and act on this fact. For Vesalius could, and did, dissect *human* corpses. Galen could not and did not. And this fact was crucial. In his dissections, as we have seen, Vesalius was gradually noting more and more points about the human bodies he was dissecting which were at odds with what Galen had said. In themselves each of these points was trivial: the insertion of a muscle here, the course of a vein there. But they had a cumulative effect to the point where there dawned on him an overwhelming thought. 'I read Galen's book *On the bones* to the students at least three times', Vesalius was later to write, 'before I dared perceive any mistake on Galen's part, although now I can't be sufficiently astonished at my own stupidity at understanding so little what was written, and thus myself deceiving my own eyes.'[37] Vesalius acquired eyes to see that the human body is what Galen was describing: but the human body is not what he had been dissecting. Vesalius acquired ears to hear that Galen himself said this: Vesalius could now understand '*what was written*' (*quae scribebantur*), what Galen himself had said. But Vesalius did have human bodies to dissect, thus his duty was now clear: to redo the whole of Galen's anatomy, but on the human body, the body it was supposedly about. In order properly to *be* Galen, Vesalius had to do what had not been available to Galen to do.

Given that Galen had said several times in his writings that force of circumstance prevented him from dissecting humans, this might not seem a particularly great insight on Vesalius's part. It was not as if he was catching out the great Ancient in a lie or subterfuge. All Vesalius was doing was taking Galen seriously, but what this now meant was that Galen's anatomy must be recognized as being the anatomy not of man but of the ape. And if it is not to be found in the writings of Galen, then nowhere in the surviving anatomical writings of the Ancients do we have an account of the anatomy of the human. The anatomy of the human, therefore, still has to be done for the Moderns for the first time. And when it is done, then and only then will we have the account of anatomy which authentically underlies the rest of Galen's medical system. These are the thoughts that spurred Vesalius on to a most astounding piece of dedicated effort, astounding because of the speed and thoroughness with which it was executed.

The idea, the 'blinding flash' as it were, occurred to him most probably some time in early 1540; By August of 1542 the entire, enormous book was finished ready for printing, with all the woodblocks cut for the most detailed, intricate and spectacular anatomical illustrations that the world had ever seen. The young man was in a hurry. Vesalius worked like a man possessed, as if only a couple of years had been allowed to him to complete this monumental task. Some of the work had, of course, already incidentally been done: Vesalius had extensive experience of dissection, and (as we have already seen) had for some time been building up a series of detailed figures of the parts of the body. But a great deal of work lay ahead. Vesalius threw all his abundant energies into the task. That task, as he conceived it, was to write a *Fabrica*: a complete account of the *structure* of the human body, as it was visible in dissection. The *actions* and *uses* of the parts were not to be omitted, but what the book was centrally about – as Vesalius's own title for it proclaims – was the *fabric*. And his guide and model in this enterprise was to be nothing other than Galen's book on dissection: *On anatomical procedures*!

Of course, the availability of human bodies to dissect was only a necessary condition for Vesalius's realization, not a sufficient one. For we have seen that anatomists had been dissecting human corpses for at least 200 years, since at least the time of Mundinus, and none of them had thought that the anatomy of Galen needed to be redone. So there is something singular about Vesalius's attitude here. To make the contrast, Vesalius's position here may be compared with that of another man also passionately concerned with Galen and anatomy, someone indeed who shared lodgings with Vesalius in Padua while Vesalius was working on his *Fabrica*. This was John Caius from England.[38] He was untouched by Vesalius's insight. Both of them were working on Galen for the 1541 Guinta edition of the *Complete Works*. Vesalius was working on the nerve and artery/vein texts for the Guinta edition, and on revising Guenther's version of *On anatomical procedures*; Caius on the Greek text of the same work, *On anatomical procedures*. They did not get on. They certainly had different views on the fallibility of Galen. Caius thought Vesalius had corrupted Galen's meaning in several places. Indeed Caius's later response to Vesalius's claims in the *Fabrica* was to assemble all the passages in Galen which seemed to show that Galen really had dissected man, and to plan an edition of Galen to defend him against Vesalius.[39] The most important event in Caius's experiences at Padua was not his close acquaintance with Vesalius, but his attendance at the lectures of Montanus on Galen's *Method of healing*.[40] Caius later returned to England where he had an honourable career as a court physician, and refounded Gonville Hall at Cambridge University.[41]

While Vesalius was writing the *Fabrica* he dissected corpses day after day; he made preparations of various parts of them, had them illustrated, and wrote the text. We do not know the names of the person or persons who produced the illustrations or made the woodcuts, although, whoever they were, we do know that Vesalius lost his temper with them.[42] This work was done in Venice, which was probably the best place in the world at that date for both artists and blockcutters. But the actual printing of the work was to be done not in Venice but, for unknown reasons, over the Alps in Basle.

Rarely can the physical realization of a printed book have been so much the brainchild of one man. Vesalius was responsible not just for the text, not just for the concept and perhaps even the initial drawing of every one of the many illustrations, but also for every detail of the printing. Vesalius was obsessed by every point about the book. There is nothing 'accidental' about it (except of course the typographical errors). Vesalius knew what he wanted to achieve – everything is judged to a hair. He wanted the book to make a tremendous impact, and it made it. Nothing is omitted which he wanted to put in, and nothing is put in which he wanted to omit.

What, then, is the structure of this enormous book? First there is an illustrated title-page, which we will be analysing later. Then comes a Preface addressed to the Holy Roman Emperor, Charles V, who was the employer of Vesalius's father and later of Vesalius himself. A letter from Vesalius to the printer is next included, from which we know about Vesalius's obsessive concern with the printed format of the *Fabrica*. A portrait of Vesalius displaying a dissection of the muscles of the inner arm (see Figure 4.12) proudly faces the first page of text. And this, like every other 'book' and 'chapter' within the work, has a specially cut initial letter; this first book is about the bones, so the initial letter shows *putti* boiling some bones in order to clean them to make a skeleton (see

4.16 *Putti* boiling the bones preparatory to making a skeleton. From the *Fabrica*, 1543, p. 1, initial letter.

Figure 4.16). And as Vesalius progresses through his presentation, at every point it matches what Galen had done in the *Anatomical procedures*. It is Vesalius's own Bologna demonstrations turned into text. As Vesalius himself says, 'In arranging the order of these books *I have followed the opinion of Galen* who considered that after the account of the muscles, the anatomy of the veins, arteries and nerves, and then of the viscera ought to be considered.'[43]

The kind of criticism Vesalius made of Galen in his Bologna lectures is continued in the *Fabrica*. Vesalius claimed that he had discovered over 300 errors in Galen, and he makes a lot of fuss about them. But they were all relatively trivial and derive, as Vesalius repeatedly says, from Galen's enforced reliance on animal dissection in order to work out the anatomy of man. In book 7, on the brain, for instance, Vesalius criticizes the reason Galen put forward for why the cerebellum is single, whereas the cerebrum is double:

> Galen . . . [thinks] that it would have been unfitting for man to have two dorsa or two dorsal marrows. Just as if there could not have proceeded from the cerebrum – which was necessarily divided into two and was not continuous in substance – one single dorsal marrow, and further as if the cerebellum had necessarily been created single so as to form an appropriate organ for the undivided dorsal marrow.
>
> But oh, Galen, you who time and again laugh at – and often even when you were unfortunately deluded by your apes – and mock Herophilus, Lykus, Andreas, and other Alexandrian professors of dissection who worked on human bodies! Behold the cerebrum in the middle of its base, as in its higher parts, is but *one* and not separate, and it sends out but *one* dorsal marrow, and on the upper side near its origin Nature has excavated a space, which we call the fourth ventricle. And this ventricle is roofed by the cerebellum which is connected at the sides, and there only, with the sides of the dorsal marrow, by two round attachments [= brachia pontis], continuous with that marrow. Along the midline of the cerebellum, neither posteriorly nor anteriorly, is any part continuous with dorsal marrow or attached to it, except only for interposition of tenuis behind. Thus has Nature, believe me, created the cerebellum the better to adapt and protect the spirit transmitted to the dorsal marrow from the ventricles.[44]

Does Vesalius see a new body, one different from that seen by his contemporaries? Most certainly yes. He has resurrected Galen's view of the *dissected* body, what I referred to earlier as *anatomy as structure*.[45] What Vesalius sees is the human body, *built on and supported by* the bones, to which the *muscles are attached*; it has *great systems of vessels* (the veins, arteries and nerves). In all these ways it differs from the 'Mundinus' body. It still has *three venters*, however, though they are now dealt with in a way which puts them into properly Galenic perspective. Vesalius sees in this body most of the parts and features Galen had seen, though he gradually learnt to doubt whether he could see the *rete mirabile* or the pores in the septum of the heart. But fundamentally what Vesalius sees is the body Galen had wanted to see, and this is new.

Hitherto, as we saw with Curtius, Galen was the text, and the demonstration of the body adhered to Galen. Now, with Vesalius, the body is the text. The body is a better text than Galen, and where Galen differs from the text of the body, then Galen must be ignored. The true text, the text of the human body, is in front of one's eyes at the demonstration, and all those attending it have a duty to read it for themselves. More than once Vesalius uses just this vocabulary of body and text. At one time he speaks about 'paradoxes drawn from *the non-lying book of the body*',[46] at another of awaiting an opportunity of dissecting '*the true book of ours, the human body* – man himself' who, because of his great abundance of remarkable things and the artifice of his Maker that he shows, is most worthy of being investigated.[47]

(iv)

More about the nature of Vesalius's enterprise can be discovered from looking at the argument of the Preface to the *Fabrica*, which Vesalius addressed to the emperor, Charles V. This will also show us where Vesalius found a more pristine, more pure form of anatomizing on which to model his own practice.

The Preface starts with a lament at the dividing up, over time, of disciplines. This is then applied to medicine, with Vesalius appealing to the history of medicine. The story he tells is of the decline of medicine from its proper state in the time of the Ancients; only now, with the work of people like himself, is it being restored. What was the cause of this decline? It was not primarily due to the decline of ancient culture and civilization; nor to the loss of learning. Rather the decline was due to people *ceasing to use the hand in medicine*. The history of medicine that Vesalius puts forward is a history of the use of the hand: when the hand is used, medicine flourishes; when it is neglected, medicine languishes; when it is restored to use, medicine can flourish again. Medicine declined 'because its primary instrument, applying the activity of the hand in curing, was so neglected'.[48] The ancient physicians each performed all aspects of medicine; they made up their own drugs, they performed operations on the body, they dealt with internal medicine. Even the members of the Rationalist, Empiric and Methodist sects of the time of Celsus and Galen, sects which may have quarrelled over everything else, all undertook all parts of medicine, the mental and the manual, surgery, pharmacy and internal medicine: 'this triple method of healing was equally familiar to the physicians of each sect'.[49] But for the last several hundred years, the use of the hand has been separated from the other aspects of medicine; indeed, learned men have come to despise working with their hands. The making-up of drugs was entrusted to ignorant apothecaries, and surgery to equally ignorant surgeons; physicians, neglecting the use of the hand, became equally ignorant of medicine as a whole. Medicine will only flourish again when practitioners again begin to use their hands, when all three parts of medicine are again reunited in the practice of every practitioner. Students of medicine ought 'to be urged in every way . . . in the Greek manner, and in the way the nature of the art thoroughly teaches, also to employ their hands in treatment'.[50] And this division of medicine has had the most dire consequences for anatomy, that 'chief branch of Natural Philosophy which, since it includes the *historia* of man, ought rightfully to be considered both the most solid foundation of the whole art of medicine, and the basis of its teaching (*constitutionis initium*)'.[51] The most important parts of anatomy (the bones, muscles, arteries, veins and nerves) have been totally neglected, and what has been taught (that is, the organs in the three venters) has been taught by the wrong people, one up aloft in his professorial *cathedra* who does not touch the body, and one below who cuts up the body, but is innocent of learning.

Vesalius himself was in a possibly unique position in his age in bridging the three aspects of medicine: he was the son of an apothecary; he was a teacher of surgery; he was qualified as a physician; and he was passionate about, and supremely skilful at, the manual business of dissection. But is this emphasis on the importance of the use of the hand anything more than a piece of self-serving rhetoric? Yes, it is more than rhetoric, and for several reasons. In the first place we have seen Vesalius engaged in trying to reinstitute not just the knowledge of the Ancients, but one of their *practices*: he himself engages in anatomizing, he *participates*, and he exhorts other people to do so too. It is not (as it was in the case of Sylvius) the practice of anatomy in order to see for oneself what Galen had written about: a sort of short-term means to a higher end. Instead, it is practice for the sake of the practice – for the sake of engaging in a practice of the Ancients. To engage in a practice, one which involves physical things, necessitates the use of the hand. And the use of one's own hand cuts out all intermediaries, it gets round that absurdity (as Vesalius saw it) of having a professor lecturing from his *cathedra*, while someone else did the dissection. Every man is responsible for his own understanding of what the human body is like. Only the use of one's own hand, and one's own eyes, will give one this authority, will allow one to reclaim this responsibility. And to employ the hand is of course simply to do what the Ancients themselves did.

The second obvious way in which this emphasis on the hand is not simply self-serving lies in the theme of the self-portrait which Vesalius put in the *Fabrica*: the dissection of the muscles of the forearm (see Figure 4.12).[52] For these are the muscles which move the hand. Now the human hand, from Aristotle's time, had been seen as the 'instrument of instruments', that is to say, the instrument which uses other instruments.[53] Man, unlike other creatures, has no defensive parts, no claws, no scales, his feet are not particularly swift to run from predators, nor does he have fur to keep him warm. Instead of all these things (said Aristotle) he has been given the hand, the instrument of instruments, by means of which he can create defences, he can make clothes, he can provide for himself in an unlimited number of ways. Together with reason and speech, the hand is one of the three ways in which man most differs from other animals. Now, when Galen was writing his book *On the use of the parts*, it was the hand with which he began his account, singing a paean of praise to it. Indeed it was these particular muscles which, in his view, presented the unanswerable argument that there was design in Nature, and that man was the high point of that design. Vesalius too started his Bologna teaching with a demonstration of these same muscles. It will be recalled that our student note-taker at the demonstration could not help commenting that 'This was indeed most beautiful to see'.[54] Thus in portraying himself as making this particular dissection, Vesalius was portraying himself *as* Galen (see Figure 4.12); and hence he was also making a statement about the existence of design in Nature and about the status of man.

Thus the use of the hand is the means of restoring the practices of the Ancients, while the manual practice of dissection reveals the hand to be most beautifully constructed by Nature – and hence something which Nature obviously intended men to use. It is the instrument of instruments, the only means by which authentically ancient practices can be revived, both in anatomy and medicine itself.

If we return now to the Preface, we can hear Vesalius saying that it is the renewed use of the hand in the present age that is leading to the resuscitation of anatomy, such that we may claim that 'the present system (*ratio*) of anatomy can be compared to the ancient one'.[55] Vesalius can claim that in the anatomizing he has conducted in Padua and Bologna, and equally here in this book, he has done things so thoroughly that 'we can desire nothing which was done by the Ancients and has come down to us, and the construction of no part which may be met with may here be wished for'.[56]

Vesalius's aim has been to restore the hand-based anatomical practice of the Ancients. Putting this programme into effect has led him to see certain shortcomings in the account that Galen offered: and then he noticed that Galen himself 'had never dissected the human body'.[57] But, he discovered, there were some Ancients who had actually done so – Ancients who lived long before Galen. Vesalius has found an ancient anatomical practice more pristine than that of Galen! It consisted in dissecting *humans* – precisely what Galen did not do. What Vesalius has discovered is the practice of Herophilus and others, and he discovers it from almost the only possible source – those places in Galen's own writings where Galen mentions, and usually criticizes, such earlier anatomists. That they were (by Celsus's account) *vivi*secting rather than just dissecting, Vesalius does not mention (but then nor had Galen); nor does he mention the fact that one of those who practised it was Erasistratus. The virtue of such anatomists in Vesalius's eyes was that the body they dissected was the human body. This is 'the true knowledge' of anatomy, 'and there is hope that it will soon be cultivated in all our Academies in the same way as it was once conducted in Alexandria in the age of Herophilus, Andreas, Marinus and other most outstanding leaders of dissection'.[58] Vesalius had started by trying to revive the anatomical practice of Galen. He has ended by calling for the revival of what was a more pristine anatomical practice: human anatomy performed on human corpses. Vesalius the Modern has become an Ancient: he has recreated an ancient practice – the practice of *human* anatomy.

(v)

All the points about Vesalius and his enterprise which I have been teasing from his biography and work, can be seen loudly and boldly announced on the title-page of the *Fabrica* itself (see Figure 4.17). This title-page is loaded with meanings: literal (historic), symbolic, allegorical. Vesalius is the person who put them there. We have no reason to doubt, and many reasons to trust, that Vesalius had as much control over this page as he did over any of the others. Every picture tells a story, and this one tells many. They are all Vesalius's stories.

Let us start by looking at the picture as a piece of theatre. This is totally appropriate for the picture is literally of a theatre, one just like (as Vesalius says)[59] those he had had created for his Paduan and Bolognan demonstrations. And first let us look at it 'historically', as a real event dramatically captured as if on a photograph. Here we can see it as portraying a demonstration such as Vesalius frequently gave, with the body of a woman opened at the first venter, an articulated skeleton

4.17 Title-page of the *Fabrica*, 1543, first edition.

available to point out the bony frame of all the muscles and other parts, the students behaving in a riotously enthusiastic way, a crowd of onlookers variously arranged, Vesalius's coat of arms (three weasels) prominently displayed, a monkey and two Venetian dogs waiting their turn to be dissected or vivisected. It is in this literal way that historians have tended to look at it hitherto. But the picture is of course theatre in a different sense too: a dramatic presentation with 'messages', morals, for the onlooker and prospective reader of the book. As they are being presented on the title-page, such messages are prominent and loud indications of what the book itself is about; they are, as it were, explanatory subtitles to the title proper. The people for whom Vesalius intended this book would have been able to read these messages. And some of them at least would have seen their novelty.

Looking at it as theatre in this way we should notice the fact that our eyes are led inexorably to identify the central actor – who is *Vesalius himself*. And, as the central actor, Vesalius is seen at the dissection table, *engaged in a performance*: he is actually doing a dissection. This is the first message, and it is about *personal engagement*. The second message is one of *changed ritual*; Vesalius is the *only* central actor: he both lectures and dissects. The *audience* for this new version of an old ritual is also changed; around Vesalius are *young* people. They are eagerly reaching forward and touching the body; they too are *participating* in this ritual. The demonstration is primarily for them, the medical students; greybeards have been relegated to the back.[60] The students are being *shown how to dissect*.

Then there are the three extra-large figures in the foreground. A glance at the sketch (see Figure 4.18) which has been identified by Ivins on technical grounds as the probable original for the woodcut,[61] will reveal that these three figures were always intended to be larger than life (and that the third, half-hidden, figure did once have his own head). Who are they? And why are they dressed differently from everyone else? That they are larger – giants – is itself a clue. That they wear sandals and seem to be dressed in sheets, is another. For they are Ancients. The one on the right, directing the gaze of the viewer into the picture, and as venerable as artistic convention said a philosopher should be, must be Aristotle, especially in this context as he is looking at animals. The one in the centre, wearing a physician's prescription case on his belt, can only be the Prince of Physicians, Galen. The identity of the third Ancient, half-hidden but the largest of all in stature, is more open to speculation. The fact that he is half-hidden by Galen might be strong grounds for identifying him as Herophilus, though it is also possible he may be Hippocrates.[62] And what are they doing? They are paying attention to, or by gesture calling attention to, Vesalius as he dissects. They are watching Vesalius resuscitating

4.18 Original sketch of title-page of the *Fabrica*, 1543. National Swedish Art Museum (C. 16.135), Stockholm.

their practice. Their presence and approval makes Vesalius one of them: he is a Modern Ancient. This is the allegorical message.

Finally, Vesalius is pointing with each hand, while looking out of the picture – an artistic convention that he is a man with a message for the viewer. At first inspection Vesalius appears to be pointing with one hand at the skeleton which towers over the dissection table, and with the other at the body being dissected. If so, then the message here would be about us remembering our own mortality: know thy mortality (the skeleton, the death-figure), remember that the live body becomes a dead body – *memento mori*. And this would have been an acceptable message in the period.[63] Indeed one of the skeleton figures of the *Fabrica* is giving a conceit of just this message (see Figure 4.19). (It is only a conceit, since the *momento mori* requires one of the two figures to be portrayed as alive, the other as dead, but both are dead in this picture.) By a strange quirk the title-page did indeed come to give this *memento mori* message.

But originally the title-page gave a different message. For, as the original sketch of the title-page (Figure 4.18) reveals, Vesalius did not initially intend there to be a skeleton at all in the picture. Hence Vesalius's gesture was one of simply pointing upwards – to God, to heaven. And thus his message was this: *personal experience of dissection of the human body, reveals God to us*.[64] But at some time before publication Vesalius decided to add to the picture a skeleton. As we know, he was accustomed to erect a skeleton over the dissection table when he was demonstrating, with an iron rod up the spine to give it rigidity. The skeleton was to be leaning on a stick, in one of the naturalistic poses that Vesalius loved. As he described it in the *Fabrica*: 'This [desired pose] you should especially seek by using a stick, by which the hands may be stabilised; a reaper's scythe gives one position, a lance another, and again a spear gives another, or a trident of Neptune, or any other instrument.'[65] But there is no reason to think that because he had now introduced a skeleton on to the scene the message therefore was meant to be about human mortality: Vesalius is still pointing upwards, at the heavens, not at the skeleton. And then when the second edition of the *Fabrica* was being prepared, and the title-page was being recut (and in a most inferior manner) without Vesalius's supervision,[66] a most unfortunate thing happened. The cutter, assuming that the skeleton was meant to be a death figure (see Figure 4.20) put a scythe-blade on the top of its stick, and a stylus into Vesalius' right hand!

If we add up these sub-title messages, we get a change in the nature of an established ritual. It is a change of personnel conducting the ritual – it is now a participatory ritual, without intermediaries; the visual rhetoric exhorts everyone to do the same; it is the resuscitation of an Ancient, a pristine or primitive, *practice*; it is a practice to do with the investigation of the human body to learn more about God and His works. Moreover, the whole event is taking place in a theatre, which may also be a church – and possibly even on the altar.[67]

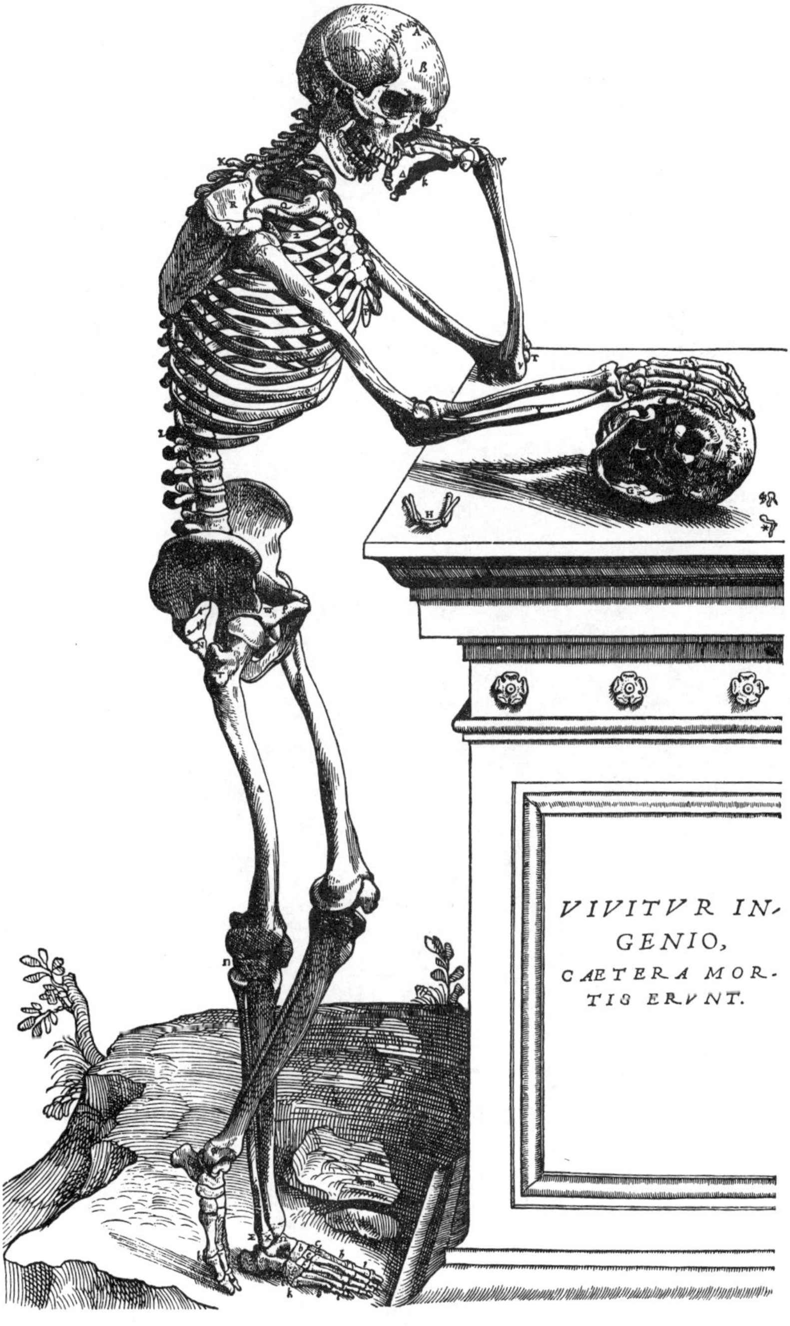

4.19 Conceit of *momento mori*: a skeleton figure from the *Fabrica*, 1543, p. 164.

4.20 Title-page of the *Fabrica*, 1555, where the skeleton has been rendered as a death figure.

(vi)

This achievement of Vesalius does indeed mark a watershed in the history of anatomizing. It marks an attempt, a successful attempt, to restore the anatomical *practice* of one of the Ancients. Anatomy was transformed thereafter: not because everyone duly copied Vesalius and complied with his exhortations, but because he had made anatomy centrally important again. Indeed he had made it exciting and controversial. As a result, there were to be full professors of anatomy in all the best medical faculties within a few years. These professors could face with equanimity the thought that the body might contain things to observe which the Ancients themselves had not seen; and, with luck, such newfound parts might be named after their discoverers. Students of medicine came to insist on seeing human anatomies, and wanted to get as close to the proceedings as possible. From now on they expected the professors to do the dissecting themselves, not just to lecture from a distant chair. Teaching anatomy from the dissection of human bodies became a central and essential part of the medical course, and (in one form or another) it has largely remained so ever since.

But Vesalius's innovation was not universally popular. Sylvius, who had been one of Vesalius's teachers in Paris as we saw earlier, found himself unable to stomach what Vesalius was claiming with respect to Galen. Hence Sylvius and Vesalius came to engage in a hostile and abusive pamphlet warfare. This was not just about particular doctrines in anatomy, but about the implications of Vesalius's stance, including issues of authority and about who has the right of specifying the authentic text and interpreting it.

Although Sylvius came to be aware that his anatomical authority – the *text* of Galen – was not totally accurate, he was so committed to it that he could not afford to have Galen corrected by (as it were) the Greek of the *body*, because that would have been to concede to Vesalius the very point at issue: the authority of Galen as teacher, and the text of Galen as the true interpretation of the body. Galen had to be protected from his calumniators. Galen had been approved by the long use of many centuries, and to Sylvius's mind, no one should dare or presume to reject Galen under any pretext whatsoever. Taking such a position in defence of Galen obliged Sylvius to maintain several impossible and unlikely things, including one that Galen himself would not have wished to maintain: that Galen had indeed anatomized humans.

The dispute is preserved mainly in three printed works. The first is Vesalius's *Letter on the china-root* of 1546.[68] Vesalius had heard that Sylvius had condemned him for claiming that Galen had never dissected humans. Vesalius wrote to Sylvius, Sylvius wrote back. In writing his

Letter on the china-root, Vesalius took the opportunity to recall this correspondence, 'running over the content of a certain letter [from himself] to Jacobus Sylvius, very useful for those interested in truth and especially in the fabric of the human – since it easily demonstrates just how much trust should be put in Galen about it', as the full title says.

> After Sylvius stated that I, Vesalius, had improperly (*indignis modis*) criticised Galen, he added what specially offended him: that I had claimed that Galen had not dissected the bodies of humans . . . he wrote that he remains firmly convinced that nothing was described wrongly by Galen. He even put it in his own letters that Galen remains beyond all blame in every instance in which I claim that he described the *historia* of the body either falsely or insufficiently correctly, or that he omitted things, or failed to give a true account of the use of the parts, or used faulty reasoning to strengthen his dogmas.[69]

The china-root occupies Vesalius for 40 pages of his book, and he then uses the other 150 to give instance after instance proving that Galen had not dissected humans but only animals, especially apes and dogs. It is quite striking how often here Vesalius presents himself as being accused of 'impiety' towards Galen.[70] The double issue constantly being raised is whether the claim is true (that Galen did not dissect humans), and whether it is impious of Vesalius to make it.

Sylvius gave a reply in print to this in a work with a punningly insulting title: *A refutation of the calumnies of a certain madman [Vaesani] against the Anatomy of Hippocrates and Galen*, in 1551.[71] Here he does indeed accuse Vesalius of impiety and of disrespect for authority and tradition, defending Galen on point after point. Here he concedes nothing to the claims of that madman, Vesalius. But in a work published only after his death, though written probably some time in the 1540s, Sylvius did make a kind of concession over whether the body as Galen describes it is or is not the same as the body one may see today in dissection. The book was *An Introduction to the Anatomic Part of the Physiology of Hippocrates and Galen*, and it was published in 1552.[72] Sylvius devoted the address to the reader to just this point. But even here, where Sylvius seems to concede the point, it is a concession turned toward proving the truth of Galen, for he argues that as the human body Galen described does not correspond in all details to the human body as we dissect it today, then the body must have changed over the centuries, due to variations in airs, waters and places over time. In other words, Galen was right, therefore it is the body which must have changed! And this decay in the physique and powers of modern man helps explain why the heroes of Antiquity were capable of such feats of strength, consumption and endurance, which are beyond modern man. 'The ancients are therefore not to be taxed with ignorance or negligence in anatomizing, but what needs to be discovered is the cause of the change in our bodies which has vitiated the nature of our bodies.'[73]

However, in 1506 there had been re-discovered in Rome a statue which had been famous in Antiquity for the perfection of its execution: that of Laocoön and his two sons being attacked by sea-serpents after warning the Trojans of the wooden horse left by the Greeks. The statue became part of the papal collection of Julius II and was regarded as an instance of the perfection of ancient sculpture (see Figure 4.21). As we have just seen, Sylvius argued that the human body had changed over the centuries and used literary evidence from Antiquity to prove his point. But in sculpture, writes the eminent historian of art H.W. Janson, 'through works of art unsurpassed for their exact rendering of anatomical detail, the ancients themselves had provided posterity with incontrovertible evidence that the structure of the human body had remained unchanged throughout the centuries'.[74] A woodcut of the Laocoön group, but with Laocoön and his sons represented as apes (see Figure 4.22), was in circulation from the mid-sixteenth century, and its significance has never been explained.[75] Janson asks what could have been a more effective pictorial rebuke to the Galenists in general and to Sylvius in particular than this *reductio ad absurdum* of the Galenist point of view, showing the most famous, as well as the most anatomically detailed, sculptural masterpiece transposed into simian terms? Janson asks:

> If we accept this as the true meaning of the woodcut, its message might be formulated as follows: 'This is what the heroic bodies of classical antiquity would have to look like in order to conform to the anatomical specifications of Galen!'[76]

Men must have been apes, if Galen's account of anatomy were right.

Sylvius was not the only person who leapt to the defence of Galen against Vesalius. For instance, John Caius also came to the defence of Galen's authority and accuracy and even tried to assemble all the references in Galen's writings which might be taken to indicate that Galen had indeed dissected humans.[77] It can be seen from such responses that both sides accepted that the claim for Galen's authority and for the authority of the tradition built on his anatomical writings stood or fell on whether Galen had indeed dissected humans. We should not here condemn Sylvius or Caius for stupidity, blindness or intransigence, or for not being able to see and appreciate the evidence of their own eyes. Both sides believed they were looking at the evidence. And it was not as if Vesalius was looking at the body, and Sylvius was only looking at books: In the opening sentence of Sylvius' address to the reader in his

4.21 Laocoön and his sons. Laocoön warned the Trojans about the wooden horse, and the goddess Athena punished him by sending sea serpents to kill him. The statue group dates from the first century BC, and was rediscovered in Rome in 1506 and placed in the Vatican.

4.22 The Laocoön group portrayed as apes. Mid-sixteenth century woodcut caricature by Boldrini, supposedly after Titian.

Isagoge, he describes the work as 'commentaries on anatomising gathered from the diligent observation (*ex diligenti observatione*) of many human bodies dissected by me in public and in private'.[78] Both parties were looking, but looking through different 'spectacles' and hence seeing different things. Hence the dispute could not simply be resolved according to some neutral 'truth of the body', because that was precisely what the dispute was about: the body seen through the 'spectacles' of Galen, who to Sylvius and Caius represented authority and tradition, compared to the body seen through the 'spectacles' of Vesalius, who to them represented dangerous impiety. It was because Vesalius claimed that 'the non-lying book of the body'[79] was superior to ancient authority and tradition as embedded in Galen's text, that he had to be opposed.

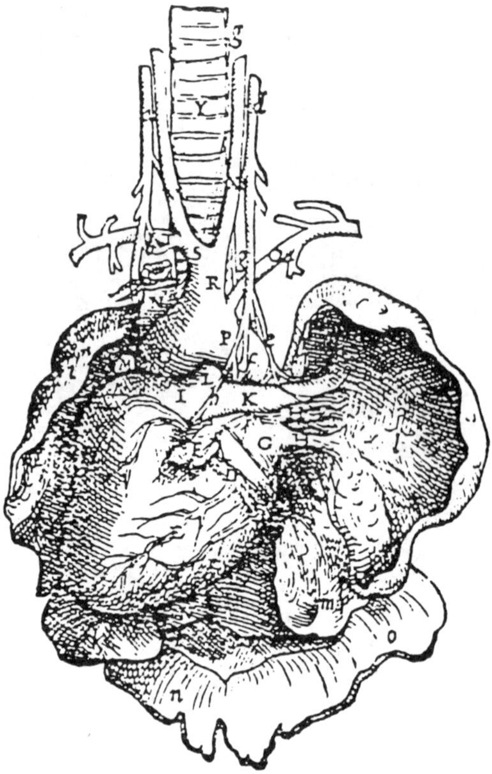

4.23 One of Vesalius's figures showing the lungs and heart of the human; note how the lungs are shown as surrounding the heart. The sixth figure of the sixth book of the *Fabrica*, 1543, p. 564.

(vi)

Finally, we need to consider briefly the vexed question of Vesalius's view of the heart, since in their search for antecedents of the discovery of the circulation of the blood, historians of anatomy have chosen to tell stories about Vesalius and other sixteenth-century anatomists which concentrate on the heart. It should be obvious that, as Vesalius was adopting the anatomical programme of Galen, he would adopt the Galenic view of the inter-relations of the organs of the middle venter. He did indeed see the heart, lung and trachea as organs united in their function of respiration. In the *Fabrica*, when he deals with the heart, Vesalius illustrates it several times as cushioned by the lungs and as constituting a set of organs with the trachea (Figure 4.23). The explanations Vesalius gives for the role and function of the various blood vessels connecting the heart and the lung is purely Galenic[80] and, as we saw with respect to the *Six Tables*, Vesalius followed Galen in seeing the blood vessels as comprised of two systems, the arteries containing *aer* or spirit in the blood to vivify the body, the veins carrying blood to nourish it. Given what Vesalius was doing – emulating Galen – there was nothing in his programme of investigation which would lead him to make any radical questioning of this view of the functions of the organs of the middle venter in the human.

Notes

1. We may take Charles O'Malley, biographer of Vesalius, as typical. In his article on Vesalius in the *Dictionary of Scientific Biography* (1976), he writes: 'The *Fabrica* was [also] written to demonstrate the fallacious character of Galenic anatomy and all that it implied . . . Vesalius regarded the *Fabrica* as the gospel of a new approach to human anatomical studies and a new method of anatomical investigation . . . The first two books [of the *Fabrica*] represent the major Vesalian achievement in terms of accuracy of description and present the most telling blows against Galenic anatomy', p. 7, cols. 1 and 2, and p. 9, col.1.
2. H.R. Trevor-Roper, *Religion, the Reformation and Social Change, and Other Essays*, London, Macmillan, 1967, p. 21.
3. C.D. O'Malley, *Andreas Vesalius of Brussels, 1514–1564*, Berkeley, University of California Press, 1965, p. 34; O'Malley's book is the chief authority on the life of Vesalius.
4. Vesalius's *Paraphrasis* on Rhazes. I rely here on the translation offered by Harvey Cushing, *A Bio-Bibliography of Andreas Vesalius*, New York, Schuman's, 1943, pp. 3–4, because my attempts to consult the work itself have not been successful.
5. Guenther, *Institutionum Anatomicarum secundum Galeni Sententiam, ad candidatos Medicinae, Libri Quatuor*, Paris, Colines, 1536, p. 32; translation based on that of O'Malley, op.cit., p. 56.

6. Vesalius writing to Fallopio in 1564, quoted in O'Malley, op.cit., p. 296.
7. William M. Ivins, Jr., 'What About the "Fabrica" of Vesalius?', in Samuel W. Lambert, Willy Wiegand and William M. Ivins Jr., *Three Vesalian Essays to Accompany the Icones Anatomicae of 1934*, New York, Macmillan, 1952, pp. 43–130.
8. For these see M. Roth, *Andreas Vesalius Bruxellensis*, Berlin, Reimer, 1892, pp. 454–57, where they are quoted.
9. My translation from Roth, op.cit., pp. 456–57.
10. This is the first of the Six Tables, which have been reprinted in facsimile in Charles Singer and C. Rabin, *A Prelude to Modern Science: Being a Discussion of the History, Sources and Circumstances of the 'Tabulae Anatomicae Sex' of Vesalius*, London, Wellcome Historical Medical Museum, 1946. The above is my translation, but see also that offered by Cushing, op.cit., pp. 12–13.
11. Ibid. The description 'honourable and beautiful, and useful and necessary' is the same as that used by Guenther.
12. On these see Cushing op.cit., pp. 16–21.
13. *Epistola, Docens Venam Axillarem Dextri Cubiti in Dolore Laterali Secandam: Et Melancholicum Succum Ex Venae Portae Ramis Ad Sedem Pertinentibus, Purgari*, Basle, Winter, p. 29. For a translation of this see John B. deC.M. Saunders and Charles Donald O'Malley, *Andreas Vesalius Bruxellensis, The Bloodletting Letter of 1539: An Annotated Translation and Study of The Evolution of Vesalius's Scientific Development* (Studies and Essays in the History of Science and Learning Offered in Homage to George Sarton on the Occasion of his Sixtieth Birthday 31 August 1944, ed. by M.F. Ashley Montague), New York, Schuman, 1946; see pp. 72–4.
14. *fomes* literally means 'tinder', that in which fire is incited; I have translated it here as 'flame'.
15. See note 13 above.
16. Saunders and O'Malley, op.cit., pp. 48–49. On the importance of witnessing, see Steven Shapin and Simon Schaffer, *Leviathan and the Air-pump: Hobbes, Boyle and the Experimental Life*, Princeton University Press, 1985.
17. This right, which obtained in Padua also, was vestigial from the origin of these universities as student-run universities. On how the students lost this right later in the century: for Bologna see Giovanna Ferrari, 'Public Anatomy Lessons and the Carnival: The Anatomy Theatre of Bologna', *Past and Present* no. 117, 1987, pp. 50–106; for Padua see Richard Palmer's forthcoming article in *History of Universities*.
18. Ruben Eriksson, *Andreas Vesalius' First Public Anatomy at Bologna 1540, An Eyewitness Report by Baldasar Heseler Medicinae Scholaris, Together with his notes on Matthaeus Curtius' Lectures on Anatomia Mundini*, Uppsala and Stockholm, Almqvist and Wiksells Boktryckeri AB, 1959. The Latin text and an English translation are on facing pages; Eriksson has edited and translated in an exemplary fashion. To bring out particular points I have offered my own version, built on that of Eriksson.
19. It is this which gives the clue to the pictures of Mundinus lecturing from his chair while a surgeon cuts and demonstrates being composites, representing in one visual space two related events which were separated in time and place. This was quite a common artistic convention. For another example see the picture of childbirth reproduced in Robert Herrlinger,

History of Medical Illustration from Antiquity to 1600, London, Pitman, 1970 p. 151.

20. On Curtius see Eriksson, op.cit., pp. 37–41; for his modernity, see Nancy Siraisi, *Avicenna in Renaissance Italy: the Canon and Medical Teaching in Italian Universities after 1500*, Princeton University Press, 1987, pp. 187–92 and the references there given.
21. O'Malley op.cit., p. 100, quoting Fabrica p. 27: 'in my last anatomy at Bologna ... prior to the dissection of the muscles, I had lectures on Galen's book On the Bones'.
22. Eriksson, op.cit., pp. 84–85.
23. *Fabrica*, p. 548.
24. Eriksson, op.cit., pp. 86–87. On Galen's employment of the three-venter division, see above Chapter 1. As Vesalius was interrupted at this point, it is not possible to know whether he was going to criticize this division or not.
25. Ibid., pp. 86-89. Eriksson identifies this as animal anatomy, see his note 13, pp. 307–08.
26. Ibid., pp. 88–89.
27. Ibid., pp. 88–89.
28. Ibid., pp. 96–97. The translation offered by William Schupbach, *The Paradox of Rembrandt's 'Anatomy of Dr Tulp'*, London, Wellcome Institute for the History of Medicine (*Medical History*, Supplement no 2), 1982, p. 16, nicely brings out the fact that these are, in modern classification, tendons: 'how the [tendons issuing from the] muscles are situated on top of each other in double formation, always four on four; and how the lower ones extend to the first joints, and the upper ones to the second and third, in each finger perforating the lower tendons'.
29. Eriksson op.cit., pp. 272–73.
30. Ibid., pp. 250–51.
31. Johannes Dryander, *Anatomia Capitis Humani, in Marpurgensi Academia Superiori Anno, Publice Exhibita*, Marburg, Cervicornus, 1536.
32. On Vesalius's work for this edition see O'Malley, op.cit., pp. 101–04.
33. Eriksson op.cit., pp. 290–91.
34. Ibid., pp. 228–29 (penis), 252–53 (lung and veins of arm), 288–89 (*rete mirabile* and head).
35. Ibid., pp. 290–93.
36. Eriksson op.cit., pp. 292–93; Vesalius gave his answer at the end of the previous demonstration: viz., that the two movements are different in time.
37. This is as translated by O'Malley op.cit., p. 111. The Latin is as follows: 'Ego ex me vel inde conjecturam facere possum, quod Galeni librum de Ossibus vel ter studiosis praelegerim, priusquam mendam aliquam Galeni annotare ausus fuissem: quum tamen nunc satis nequeam mirari meam stupiditatem, qui, quae scribebantur, tam parum assequerer, et ipse meis oculis ita imponerem', *Andreae Vesalii Bruxellenisis, Medici Caesarei Epistola, Rationem Modumque Propinandi Radicis Chynae Decocti ... Pertractans*, Basle, 1566, p. 196.
38. See O'Malley op.cit., p. 101: the precise date is unknown.
39. See O'Malley op.cit., p. 107. On Caius's reactions in defence of Galen, see Vivian Nutton, '"Prisci Dissectionum Professores": Greek Texts and Renaissance Anatomists', in A.C. Dionisotti, Anthony Grafton and Jill

Kraye, *The Uses of Greek and Latin: Historical Essays*, London, Warburg Institute, 1988, pp. 111–26, esp. pp. 119–21.

40. See above, Chapter 3. On the (cordial) relations of Vesalius and Montanus, see Vivian Nutton, 'Montanus, Vesalius and the haemorrhoidal veins', *Clio Medica*, 1983, **18**, pp. 33–36.
41. On Caius's life and work see the Memoir of his life by John Venn, in E.S. Roberts (ed.), *The Works of John Caius, M.D.*, C.U.P., 1912, and Vivian Nutton, *John Caius and the Manuscripts of Galen*, Cambridge Philological Society, 1987. It should be noted that Caius was instrumental in acquiring the right to dissect two human bodies for anatomical demonstration at Caius College, Cambridge, in 1564 in connection with his refoundation of it; see Venn's Memoir, p. 29, in Roberts, op.cit.
42. See O'Malley op.cit., p. 124. For a possible chalk sketch preliminary to the woodcuts, see Martin Kemp, 'A Drawing for the Fabrica; and Some Thoughts upon the Vesalian Muscle-men' *Medical History*, 1970, **14**, pp. 277–88.
43. My emphasis; *Fabrica* f. 3v–4r.
44. Modified from the translation offered in Charles Singer, *Vesalius on the Human Brain*, Oxford, Oxford University Press, for Wellcome Historical Medical Museum, 1952, p. 27.
45. See above, Chapter 1.
46. '*paradoxa ex hominis haud mendaci libro deprompta sunt*'. This was found, on p. 600 of the Fabrica, by Ludwig Edelstein, 'Andreas Vesalius, the Humanist', *Bulletin of the History of Medicine*, 1943, **14**, pp. 547–61, reprinted in Owsei Temkin and C. Lilian Temkin, *Ancient Medicine: Selected Papers of Ludwig Edelstein*, Baltimore, Johns Hopkins Press, 1967, pp. 441–54, see p. 449. Edelstein says that the term seems original with Vesalius.
47. '*verum illum nostrum humani corporis librum*'; from *Andreae Vesalii, Anatomicarum Gabrielis Falloppii Observationum Examen*, Venice, 1564, p. 171.
48. My translation from *Fabrica* f. 2r. A translation of the Preface is offered by O'Malley, op.cit., Appendix, pp. 317–24.
49. *Fabrica* f. 2r.
50. *Fabrica* f. 2v.
51. *Fabrica* f. 2v.
52. Very important here is the outstanding study by William Schupbach, *The Paradox of Rembrandt's 'Anatomy of Dr Tulp'*, above note 28. Though I differ from Schupbach on some issues, I depend on his discussion.
53. See Schupbach op.cit., Appendix II, pp. 57–65, 'The special significance of the hand', for all the pertinent passages from Aristotle.
54. For Vesalius on the hand, see Schupbach op.cit., pp. 18 and 59, citing and translating *Fabrica* p. 305.
55. *Fabrica* f. 3r.
56. *Fabrica* f. 3r.
57. *Fabrica* f. 3v.
58. *Fabrica* f. 3v.
59. *Fabrica* p. 548.
60. On the variety of people portrayed as present, and on their dress, see Giovanna Ferrari, 'Public Anatomy Lessons and the Carnival: The Anatomy Theatre of Bologna', *Past and Present* no. 117, **1987**, pp. 50–106.

61. See William M. Ivins, Jr., 'What about the "Fabrica" of Vesalius?', in Samuel W. Lambert, Willy Wiegand and William M. Ivins, Jr., *Three Vesalian Essays to Accompany the Icones Anatomicae of 1934*, New York, Macmillan, 1952, pp. 43–130.
62. Cushing, op.cit., pp. 81–82 identifies the man on the right as Realdus Columbus 'motioning a servant to quiet the dog'; Castiglioni, op.cit., p. 835 thought this same character represented Galen, but 'very dissatisfied with the whole proceedings, looking away with manifest disdain'. O'Malley, op.cit., p. 144, thought the two forward characters (he did not notice the third one) were Vesalius's patrons.
63. See Schupbach, op.cit., for many instances of this. Schupbach's whole theme is that 'at some time in the early sixteenth century a new intellectual fashion had appeared among anatomists: the invocation of the Greek proverb *gnothi seauton, nosce ipsum*, 'know thyself', to justify the study of anatomy' (p. 31). Though he says that 'the origin of this anatomical application of the phrase is obscure', he suggests that it arose in the 1530s either in Venice or Padua or Paris or Montpellier (pp. 31, 66).
64. On the interpretation of gesture in painting of this period see Michael Baxendall, *Painting and Experience in Fifteenth Century Italy*, Oxford University Press, 1972; repr. 1988. On p. 65 of this work Baxandall quotes a list of 'histrionic accents', published in English in 1520: '(1) . . . when thou spekest of a solempne mater to stande up ryghte with lytell mevynge of thy body, but poyntynge it with thy fore fynger . . . (3) And when thou spekyst of any hevenly or godly thynges to loke up and pointe towards the skye with thy finger'. Both these histrionic accents can be observed in the pointing of Vesalius in this Frontispiece.
65. *Fabrica*, p. 160, my translation.
66. See O'Malley, op.cit., p. 271.
67. It is certainly the case that late sixteenth and seventeenth century anatomical theatres in Protestant regions were often in churches and the dissection was performed on the altar. See William S. Heckscher, *Rembrandt's Anatomy of Dr Tulp*, New York, 1958, pp. 29–30.
68. *Andreae Vesalii Bruxellensis, Medici Caesari Epistola, Rationem Modumque Propinandi Radicis Chynae Decocti, Quo Nuper Invictissimus Carolus V Imperator Usus Est, Pertractans: Et Praeter Alia Quaedam, Epistolae Cuiusdam Ad Iacobum Sylvium Sententiam Recensens, Veritatis Ac Potissimum Humanae Fabricae Studiosis Perutilem: Quum Qui Hactenus In Illa Nimium Galeno Creditum Sit, Facile Commonstret*, Basle, Oporinus, 1546.
69. Ibid., pp. 42–43, my translation; cf. O'Malley, op.cit., p. 219.
70. Ibid., pp. 71, 95; the point about the frequency of the use of the term 'impiety' by Sylvius, and his contrasting it with his own intended 'pious' reader, is made by Roger French, 'Natural philosophy and Anatomy', in J. Céard, M.M. Fontaine, J.C. Margolin, eds, *Le Corps à la Renaissance*, Paris, 1990, 447–60.
71. *Vaesani cuiusdam Calumniarum in Hippocratis Galenique Rem Anatomicam Depulsio*, Paris, 1555; I have used the Venice 1556 edition.
72. *In Hippocratis et Galeni Physiologiae Partem Anatomicam Isagoge*, Paris 1552; I have used the Venice 1556 edition.
73. Ibid., folio 11 r, my translation. See also the letter in *The Times* of London from R.K. French, October 8 1980, 'Golden age of drinking'.

74. H.W. Janson, 'Titian's *Laocoön Caricature* and the Vesalian–Galenist Controversy', *The Art Bulletin*, 1946, 28, pp. 49–53, see p. 51.
75. As Janson says, it has traditionally been attributed to Titian or his workshop.
76. Jenson attributes the idea for the woodcut to Vesalius.
77. On Caius's defensive reactions, see Vivian Nutton '"Prisci Dissectionum Professores"', note 39 above.
78. My translation.
79. See note 46 above.
80. See *Fabrica*, pp. 569–70.

CHAPTER FIVE

Columbus: The Revival of Alexandrian Anatomy

Today Realdus Columbus (Realdo Colombo, *c.* 1515–1559) is usually only celebrated by historians as a contender for first discoverer of the pulmonary transit of the blood,[1] and his role in reviving the anatomical tradition of Herophilus and Erasistratus has been forgotten. Remarkably little is known directly about the biography of Columbus,[2] but something can be inferred about his views from the fact of his close friendship with a very famous man, Michelangelo, with whom Columbus planned to write and illustrate a work on anatomy.

Columbus came from Cremona. Son of an apothecary, he studied philosophy at Milan, and then for several years trained to be a surgeon under a Venetian, Giovanni Antonio Plato, known as Lonigo (Leonicus), who in 1536 gave a course of anatomical demonstrations at Padua. By 1538 Columbus too had arrived at Padua, with ambitions to become a physician. While studying medicine here he was able to lecture to arts students on sophistics (logic). He also gave special attention to anatomy. These were the years of Vesalius at Padua. Columbus became a close friend of Vesalius (Vesalius referred to him in the *Fabrica* as 'mihi admodum familiaris'). Possibly he assisted Vesalius at dissection, though it is not known whether he was in any formal sense a pupil of Vesalius. In 1541 Columbus seems to have been chosen for the second of the two posts in surgery held until then by Vesalius. While Vesalius was away in Basle overseeing the printing of the *Fabrica*, Columbus was appointed temporarily to teach in Vesalius's place, and when Vesalius took the appointment with the emperor, Columbus received the Paduan position more permanently.

For reasons unknown to us, Vesalius and Columbus fell out. They were friends enough up to the time Vesalius published the *Fabrica*, for that is where Vesalius speaks of their close friendship. When substituting for Vesalius in Padua, Columbus seems to have criticized Vesalius's views in some way – apparently by claiming to have found a vein that Vesalius said did not exist – and Vesalius thereafter referred scathingly to Columbus as 'an uneducated man, a half-knower', and scorned his views on anatomy.[3]

In 1544 Columbus was enticed to the *Studio* (or university) of Pisa, which had just been reopened under the patronage and control of the

Duke Cosimo I de' Medici, ruler of Florence, who was trying to attract well-qualified teachers.[4] As Master of Anatomy and Surgery there[5] Columbus performed several dissections; he was able to use a text 'at his own discretion' (*prout eis*).[6] But then in 1547 for a while, and in 1548 for good, Columbus went to Rome. In Rome he henceforth taught anatomy at the papal university refounded by Leo X, the *Sapienza*, presumably appointed to this position by Pope Paul III.[7] His position at Pisa was given to Gabriel Fallopius. Columbus was now in the heart of Counter-Reformation Rome, in the university founded and patronized by the popes. Here he taught anatomy for about ten years, until his death.

Columbus came to Rome in order to engage in a project with Michelangelo, the pope's favourite sculptor, painter and architect, who was then in his 70s. According to the letter Columbus wrote to Duke Cosimo in 1548, requesting permission to go, this project was a work on anatomy which would give students

> the truth of the matter, and which may be studied in conjunction with dissection. I know the damage they suffer from studying Galen who, besides being lengthy, is also in error; similarly Vesalius is prolix and suffers from not a few errors, as I have demonstrated in public and as I am about to demonstrate more widely if Your Excellency will not prevent me.[8]

In Rome, Columbus wrote, there was an ample supply of bodies, necessary in order to have evidence 'to contradict the ancients and moderns alike' ('*da contradire e alli antiqui et alli moderni piu*'). The work has already progressed:

> When I finished teaching the anatomy course [at Pisa] last year, I returned to Rome with Your good leave . . . to pursue my dissections and supervise the painters. At this writing a good deal has already been done, and this coming summer we will attend to the bones.[9]

It has been generally assumed by historians that the book on anatomy that Columbus finished, and that was published just after his death in 1559, the *De re anatomica Libri XV*, is the result of the work undertaken with Michelangelo. Certainly Columbus and Michelangelo remained close for years; Michelangelo's contemporary biographer says Columbus was 'a very particular friend' of Michelangelo, as well as his personal physician.[10] But the *De re anatomica* of Columbus is probably not the result of this collaboration, and should therefore be read as a work in an anatomical tradition rather than in an art-history/anatomical tradition. To appreciate this we need first to consider Michelangelo's plan with respect to anatomy.

Both Condivi and Vasari, who gave admiring accounts of Michelangelo while he was still alive (Vasari first in 1550, Condivi in 1551–2), and who were concerned with representing him as some kind of superhuman, made the case that he had a very strong and close interest in anatomy. He was for a long time thought to have sketched a dissection scene (see Figure 5.1).[11] From before the age of 20 (that is, before 1495), according to Condivi, Michelangelo applied himself to the study of anatomy: the prior of St. Spirito 'provided him with a room and with corpses for the study of anatomy, than which nothing could have given him greater pleasure. This was the first time that he applied himself to this study, and he pursued it as long as he had an opportunity.'[12] It is noteworthy that among the many nude figures in Michelangelo's portrayal in the Sistine Chapel of the The Last Judgment, there is a supposed self-portrait – as a flayed human skin.[13] Of Michelangelo's practical knowledge of anatomy, Condivi writes:

> From his boyhood Michelangelo has been a very hard worker, and to his natural gifts he has added learning, which he was determined to acquire not through the efforts and industry of others but from nature herself, which he set before him as a true example. Thus there is no animal whose anatomy he would not dissect, and he worked on so many human anatomies that those who have spent their lives at it and made it their profession hardly know as much as he does. I am speaking of the knowledge which is necessary to the art of painting and sculpture and not of the other minutiae which anatomists observe. And that this is so is demonstrated by his figures, in which there is such a concentration of art and learning that they are almost impossible for any painter whatever to imitate.[14]

And this is what Condivi wrote about Michelangelo's proposed book on anatomy:

> Now, to return to the subject of anatomy, he gave up dissecting corpses because his long handling of them had so affected his stomach that he could neither eat nor drink salutarily. It is quite true that when he gave it up he was so learned and rich in knowledge of that science that he has often had it in mind to write a treatise, as a service to those who want to work in sculpture and painting, on all manner of human movements and appearances and on the bone structure, with a brilliant theory which he arrived at through long experience. And he would have done it had he not doubted his powers and whether they were adequate to treat the subject properly and in detail, as someone would who was trained in the sciences and in exposition. I know very well that, when he reads Albrecht Dürer, he finds his work very weak, seeing in his mind how much more beautiful and useful in the study of this subject his own conception would have been. And, to tell the truth, Albrecht discusses only the proportions and varieties of human bodies, for which no fixed rule can be given, and he forms his figures straight upright like poles; as to what was more important, the movements and gestures of human beings, he says not a word.

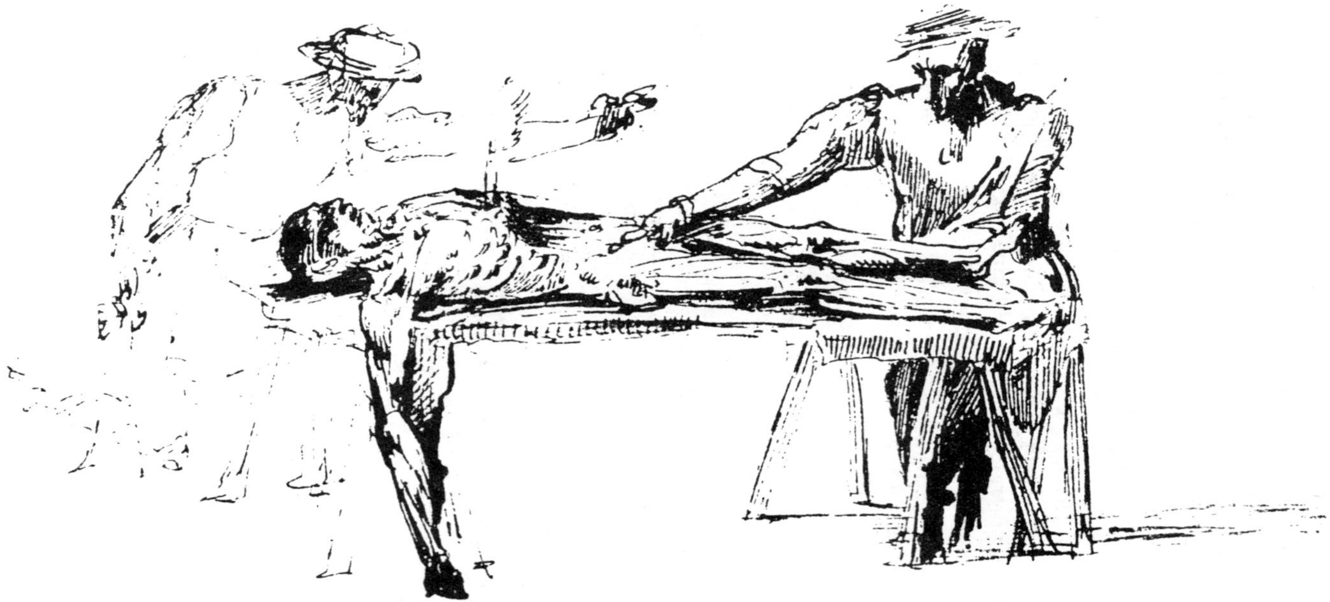

5.1 Possible dissection scene, traditionally attributed to Michelangelo, now credited to Capitelli.

> And, because by now [i.e. 1551–2] Michelangelo has attained a grave and mature age and does not expect to be able to reveal this invention of his to the world in writing, he has disclosed everything to me with great devotion and in the most minute detail. He also began to discuss this with Messer Realdo Colombo, a very superior anatomist and surgeon and a particular friend of Michelangelo's and mine, who sent him for this purpose the corpse of a Moor, a most handsome young man and, insofar as one could say, most suitable; and it was placed in S. Agata where I was and am still living, because of its being a remote place. On this corpse Michelangelo showed me many rare and recondite things, perhaps never before understood, all of which I noted and hope one day to publish with the help of some learned man for the convenience and use of all who want to work in painting and sculpture. But enough of this.[15]

This is clearly a devotion to anatomy and anatomizing in the cause of *disegno*: of painting and sculpture, and indeed also of architecture: 'I am speaking of the knowledge which is necessary to the art of painting and sculpture and not of the other minutiae which anatomists observe.' It is concerned with Michelangelo's obsession with the portrayal of the nude human figure, both in painting and in sculpture. Columbus definitely did go to Rome in order to assist Michelangelo with this project, including the supervision of the painters. But nothing is apparently known to have survived of this project, and Columbus's published book, as we shall see below, had different aims and scope and thus is not the text originally meant to accompany this projected theory and series of illustrations by Michelangelo and intended for artists.

(ii)

Columbus's own anatomical project – the one he did publish on – was concerned with his goal of restoring the practice of the early Alexandrian anatomists, Herophilus and most particularly Erasistratus. This may sound somewhat paradoxical, since Vesalius had claimed that returning to the anatomical practice of the early Alexandrians was precisely what he himself was doing, as we discussed earlier. But in Columbus's eyes Vesalius had not accomplished this but merely called for it. The key feature of such a project for Columbus was to put *vivisection* centre-stage in the search to discover more about the *actions* of the parts of the human body. Vesalius, as we will remember, was concerned primarily with giving a full account of the *fabric* of the human body, rather than with finding out more about its actions. One of the implications of Columbus's project was that it meant going back beyond Galen; it was to go back to 'the Anatomists', familiarly so called. For just as Galen was commonly termed 'the Physician' and his followers were called 'Physicians' (*Medici*), and Aristotle was 'the Philosopher' and his followers 'Philosophers' (*Philosophi*), so Herophilus and Erasistratus were referred to as 'the Anatomists'. Columbus was the first of the Moderns to see himself as, in this sense, an *Anatomicus*.[16] But the most significant implication of adopting this project of recovering the practice of these ancient Anatomists, was that it involved in many important matters not an emulation but a *rejection* of Galen in anatomy: doctrines, practice and all. For it had been against the views of Erasistratus above all that Galen had written many of his books and established his account of anatomy and set up his experimental demonstrations. Yet it was some of Erasistratus's views that Columbus was now bringing back on to the anatomical agenda.

Columbus completed the text of his book, but died shortly before it was published in 1559, and it was brought to press by his two sons, Lazarus and Phoebus. They say that Columbus had finished the 15 books of the work some years earlier and had intended to publish it soon; the books were actually being set in type at Venice when death snatched Columbus away.[17] Thus we may take the text as being authentically by Columbus, and we also know that Columbus himself had at least some say in the physical format in which it appeared. But unlike the case of Vesalius, we cannot say with certainty that the illustrated title-page – the only illustration in the whole volume – is precisely what Columbus wished to appear. Consideration of it nevertheless shows certain contrasts to that of Vesalius (see Figure 5.2). In the first place it will be impossible not to notice that this title-page is in no way as accomplished as that of the *Fabrica*; it is most unlikely therefore that Michelangelo had anything to do with it. It is a hack piece with no pretensions to being high art.[18] But in the second place we may note that it portrays a different kind of dissection event. Yes, there is only one central figure, teaching and dissecting, as there was in the Vesalian scene, and like the Vesalius title-page this one also portrays a corpse with the lowest venter opened up. But this is not a tiered 'theatre' in the Vesalian manner, only a simple table around which the attenders have gathered. Equally this is not a presentation of Columbus as a modern Ancient in the company of Ancients, as was that of Vesalius. At the front of the scene there is something as remarkable as the three Ancients of the *Fabrica* title-page: a small child, holding the hand of an old man who is looking down at the child and in the act of touching his own left shoulder region with his right hand. It is most likely that the child, in its antique dress, is supposed to represent the soul of the corpse. This has been a very familiar convention since classical times, with the departure of life often being portrayed as the child-soul exiting from the dead

REALDI COLVMBI
CREMONENSIS,
In almo Gymnasio Romano
Anatomici celeberrimi,
DE RE ANATOMICA
LIBRI XV.

VENETIIS, Ex Typographia Nicolai Beuilacquæ, M D LIX.

5.2 Title-page of *De Re Anatomica Libri XV*, by Realdo Columbo, Venice, 1559.

5.3 The soul portrayed as a little child, exiting from the body at death. From a manuscript of Aristotle, *De Anima*, 'On the Soul'. Ms. Ee.2.31, Cambridge University Library.

body (see Figure 5.3). While the man holding the hand of the child may possibly be meant to represent the guardian angel of the corpse, it is more probable that he is shown in the act of crossing himself – in the name of the Father, Son and Holy Ghost – for the sake of the spiritual welfare of the immortal soul of the deceased.

The work as a whole was dedicated by Columbus's sons to the newly elected pope, Pius IV, who had himself studied medicine in his youth.[19] The title of the work, given to it as far as we know by Columbus himself, was 'Fifteen books *de re anatomica*', literally 'On the anatomical business', and which we shall translate here as *On anatomizing* (rather than *On anatomy*) in order to convey the different scope and intention that it expresses by comparison with Vesalius's *On the fabric of the human body*. Columbus's book is primarily about dissecting, where Vesalius's had been about the dissected. As for Columbus's stated intentions, he expressed them in his address to his Candid Reader thus:

> When it occurred to me, after the very long and almost infinite labours which I undertook in dissecting the cadavers of men, to write down what I have observed and learnt by experience to agree with the nature of the thing firstly about the whole of Anatomy and secondly about dissection,[20] I thought that there would be no shortage of people who would spurn these my attempts as useless and redundant – indeed who, against those publishing novelties about anatomical things, would with great disdain counterpose their Avicenna (prince of all the disciplines, as they claim), Mundinus, Carpus – leaders in Anatomy, to whose writings on this subject it seems that nothing at all can be added. They bring out other writers: not only those I have mentioned, but Galen and Vesalius. They think that after the writings of these men, it is the act of an arrogant and vainglorious writer to deal with dissecting human bodies.
>
> But not one of these people could deter me from writing, since one can easily and truly respond to their quite frivolous objections. As far as concerns the three authors they first counterposed, I think that this familiar verse suits them:
>
> Some things are good;
Some are mediocre;
More are bad.
>
> About Galen, by contrast: I have not denied that he has been to me and other people the leader not only in Anatomy but in the other parts of medicine, and that the amount all physicians except Hippocrates owe to this great man is infinite; yet since he cut up apes and dog-faced baboons in place of human bodies, Galen's books on anatomy cannot but abound with many errors. Now, about Vesalius I will preface this: that I have always mentioned him with respect in private and in public, and said that his writings should not under any pretext be wrested from the hands of the learned; indeed by his own efforts Vesalius added many things about anatomy to the divine labours of Galen on anatomising, he was the first to detect and correct many errors in them. Yet such is the difficulty of this most noble, highly useful, and so necessary anatomic faculty, that all things can not be penetrated by one man – such that not even the huge and important volume of Vesalius *On the Fabric of the Human Body* is devoid of its own errors.
>
> For which reason, dear Reader, it should not be surprising to you if, after so many attempts of the most excellent men, I am not afraid to deal with the same matters. For since all arts and disciplines may be perfected later through the contribution of posterity, so it will doubtless be the case that after many very diligent writers on this subject, the perfected knowledge should evade even the Anatomists too; nor that my own industry in these same things can be judged superfluous. I pray therefore that you will not make a judgement about these things until you have read them all through.[21]

This was dated Rome, 1 June 1559, so it would seem that Columbus was still alive at that date. We can perhaps hear in this address to the reader a completion of Columbus's intention when he first went to Rome of finding anatomical evidence 'to contradict the ancients and moderns alike', the Moderns obviously including Vesalius and the Ancients including the great Galen himself.

We turn now to the structure of Columbus's book *On anatomizing*. It is critically important for our purposes here to look at this structure, since there is good reason to believe that it is not just something Columbus adopted for convenience of presentation, but that it is a faithful representation of his approach to anatomizing. *On anatomizing* is an anatomical text of a quite new kind, and is a momentous divergence from both the 'three-venter' tradition stemming from Mundinus, and the Galenic approach (from *On anatomical procedures*) which we have seen Vesalius to have been recently reviving, and which he used in the *Fabrica*. This can be seen particularly clearly by tabulating the contents of Columbus's and Vesalius's works against each other:

VESALIUS		COLUMBUS
Bones and skeleton	Book 1	Bones
Ligaments and muscles	Book 2	Cartilages
Veins and arteries	Book 3	Ligaments
Nerves	Book 4	Skeleton
Organs of nutrition and generation	Book 5	Muscles
Heart and organs serving it	Book 6	Liver and veins
Brain and sense organs [inc vivisection]	Book 7	Heart and arteries
	Book 8	Brain and nerves
	Book 9	Glands
	Book 10	The eyes
	Book 11	The viscera

Book 12	Formation of the fetus
Book 13	Coverings of the human body
Book 14	Vivisection
Book 15	Things rarely seen in anatomy

At first sight Columbus's arrangement, though more sub-divided and hence in more books, looks similar to the Galen–Vesalius approach, for the contents of his first five books cover the parts of the body dealt with in Vesalius's first two books. But then comes the great difference. For where Vesalius deals with the different vessels of the body in books 3 and 4 (Veins and arteries; Nerves), and then with the three great organs of the traditional 'three venters' in books 5 to 7, Columbus does something quite different. His book 6 is on the liver and veins, book 7 on the heart and arteries, and book 8 on the brain and nerves. In other words, Columbus deals with *each of these major organs together with their vessels*, where Vesalius and other anatomists dealt with these organs separately from their systems of vessels. Much was to follow from this particular choice of a way of conceptualizing the body to be anatomized, including Columbus's discovery of the pulmonary transit of the blood, as we shall see.

Then there is the matter of the lung. In earlier chapters I have made much of the fact that all anatomists after Galen (with the single exception of Columbus's exact contemporary, Servetus)[22] had dealt with the heart and lung together; they had seen the trachea, the lung, the left side of the heart, and the arteries as a set of related organs, all involved with the process of respiration. Columbus, however, has chosen to deal with the heart and arteries *but without treating the lung*. Where, then, does he deal with the lung, and in what context? The answer is in Book 11, as one of the viscera. The term 'viscera' had traditionally covered all the internal organs of the body. But Columbus uses it in a more limited way: to embrace all the organs of the body which are left after separating off the noble organs – the liver, heart and brain (and eyes) – for separate investigation.[23] By severing the conceptual link between the heart and the lung Columbus has here broken with the whole tradition of Latin anatomizing since Galen.[24] For the 'three venters' are not represented in Columbus's scheme of anatomizing. Nor is the diaphragm treated by him as being a significant natural physical boundary in the body, separating the regions of the natural and the vital organs (instead he sees it simply as a muscle).[25] And instead of being just one of the organs of the unitary process of respiration, the lung is given a special new role by Columbus. No one had tried to look at the body in quite this way since before the time of Galen.

It is as a consequence – presumably unintended – of separating his consideration of the heart from that of the lung, that Columbus came to make the discovery that the blood goes from the right side of the heart to the lung, and then returns to the left side of the heart (the pulmonary transit of the blood). He made this separation between heart and lung as a consequence of his adoption of the method of considering each of the three great organs *together with their vessels*.[26] It is therefore important for us to understand what this approach entailed, and to seek to discover why Columbus should have adopted it.

(iii)

As we will recall, it was Columbus's ambition, when he went to Rome, to acquire evidence 'to contradict the ancients and moderns alike'. In the text of *On anatomizing*, while he acknowledges their greatness, yet Columbus criticizes the most important Ancients and Moderns directly and repeatedly. He criticizes Aristotle on many particular points. Galen he criticizes for being deluded by his apes (*suis delusus simiis*),[27] and thus making false inferences about the anatomy of the human. In particular he criticizes Galen for attacking the Ancients who *had* dissected humans, while he himself had not. Those Ancients, of course, are pre-eminently Herophilus and Erasistratus at ancient Alexandria. Among the Moderns, he criticizes Vesalius for claiming to have corrected Galen and yet still showing in public demonstrations and in the illustrations of his book the anatomy of animals instead of that of the human: As Columbus says, Vesalius 'did this in public anatomical theatres, at which I was frequently present'.[28] The only other Modern whom Columbus criticizes by name is a Lutheran, Leonhard Fuchs.

Columbus's general criticism of other people is for putting reason above the evidence of the senses. It sounds very sweet in modern-day ears when Columbus writes: 'in anatomical matters Galen and Vesalius, whom I hold in the highest esteem, I do not hold as highly as I hold truth itself: I call it truth when what one says (*oratio*) agrees most closely with the nature of the thing'.[29] As evidence of the superiority of his own approach in this respect, Columbus is constantly announcing anatomical discoveries he has made himself – 'I was the first to see', 'I saw it with these eyes' – and pleading with his reader to compare his claims against the evidence of the vivisected body: 'honest Reader, I beg of you to try (*experire*) in brute animals, which you should dissect alive, I advise you and exhort you; try, I say, whether what I have said agrees with the thing itself'.[30] Yet among the many discoveries made by Columbus, thus privileging observation and sense-experience over reason, are many we no longer believe to be facts or evident to sense-experience: such as the existence of a *rete mirabile* in the human; that the brain has a contraction and expansion like the heart; and that the true site of generation of the animal spirits can be located.[31] So in itself the putting of the evidence of the senses above that of reason, important though it was, was not enough to enable Columbus to see a 'new' body.

The crucial difference in approach was that Columbus actually did what Vesalius said one should do but had not himself done: based his anatomical knowledge on the frequent performance of vivisection. In other words, Columbus anatomized the live, active body, where Vesalius had anatomized the dead body. It was this concentration on vivisection, this reviving of the practice of the ancient Alexandrian anatomists, that led Columbus to adopt his new way of conceptualizing the body: as being nourished, being vivified, and being given motion, by three life-giving rivers. There are two key passages, dealing respectively with the three 'rivers' and with the centrality of vivisection for the anatomist's enterprise, which will enable us to appreciate the unique identity of Columbus's novel, yet most Ancient, approach.

The imagery of the three rivers is most evident at the very opening of Book XI 'On the viscera':

> Although the Liver, the Heart and the Brain are viscera, and it would seem appropriate to deal with them in a treatise on the Viscera, yet I will not deal with them in this place, since I have dealt with them separately because they are like three fountain-heads from which different rivers spring (*instar trium fontium, à quibus varii proficiscuntur rivuli*), which irrigate the whole fabric of man, and without which man's fabric would be stone-like and useless. And, so that the food may flow through the individual parts of the body the veins were produced, which carry the natural blood prepared in the liver, as in its proper fountain-head, which flows along to the higher and to the lower parts . . . Similarly, from the heart is distributed the vital blood together with the heat to all parts of the body through the arteries. In the same way the brain transmits sense and motion through the nerves and spinal medulla to the whole body.[32]

There are three fountain-heads, the liver, heart and brain, from which are distributed throughout the body the three rivers of the natural blood, the vital blood (that is, natural blood imbued with vital spirits), and the animal spirits respectively. Although few earlier anatomists would have had much objection to this characterization of these three organs and the flow in their respective vessels, no earlier anatomist had made this concept of flowing rivers the practical basis of their anatomical work. Quite remarkably, this view of three rivers does not, in this precise form, come from any known ancient source.[33]

The most significant thing that this way of proceeding enabled Columbus to see concerned the lung. It made it possible for him to see, or 'discover', the pulmonary transit of the blood: that is, the passage of the blood from the right side of the heart to the left via the lung. For Columbus's way of proceeding – which is preserved within the formal structure of the book itself – was to follow the course of each of the rivers arising from each of these major organs; in other words, to trace out the route of each of the sets of vessels from their source-organs. In doing this for the heart and arteries he noticed that there is one – and only one – organ in the body which does not receive an artery from the heart. That organ is the lung. It is a large organ, and yet no arterial spur comes toward it from the aorta. In the traditional understanding (which Columbus follows here), the heart produces, and the arteries convey, blood imbued with *vital spirits* (that is, vital blood), and the presence of vital spirits in the arteries is evident from the fact of their pulsation. Any organ not in receipt of vital spirits will not be alive, will not be endowed with vitality. Yet as the lung obviously is endowed with vitality, yet it has no artery bringing vital spirits to it, Columbus concludes that it must be in receipt of vital spirits by being the *site of production* of the vital spirits!

> Here is a remarkable thing to note: that no artery tends toward the lung, which might take up the vital spirits from the heart to deliver them to the lung. What is more, this is a true axiom in anatomising: if you except the lung, the remaining parts of the human body are endowed with pulsatile arteries. On this ground, anyone acute and sharp in disputing could prove that the vital spirits are generated not even in the heart itself.[34]

The vital spirits are generated in the lung and perfected in the heart. Therefore, in Columbus's view, the natural blood comes from the liver up the vena cava. Some of it enters the right ventricle of the heart and from thence passes to the lung, along the vein-which-looks-like-an-artery; as the capacity of this vessel reveals, this amount is more than is required to nourish the lung.[35] In the lung the natural blood is mixed with vital spirits from the air, and thus the vital blood is generated. This is passed from the lung to the left ventricle along the artery-which-looks-like-a-vein; in the left ventricle the vital blood is 'perfected', and from the left ventricle it is passed to the aorta and then distributed through the arteries to all parts of the body. He does not identify the vein-which-looks-like-an-artery as an artery, since it does not carry vital blood, given that the lung is the place of generation of vital blood. This is the pulmonary transit of the blood, the 'new use of the lung' discovered by Columbus. It is a new body, revealed to Columbus by his concern to trace out the three rivers of life from their fountain-heads.

In the new body that Columbus sees, the supreme organ is the brain. Certainly the liver (which he believed to be the first organ generated in the fetus) is, as he says, 'like a king in the abdomen', and the heart 'can certainly be numbered amongst the principal parts'; but for Columbus the brain 'is more noble than all the other members, and as I might say, is the King of the principal members' of the body.[36] Obviously it was not new to speak of a hierarchy among the principal organs, for anatomists had been doing this since the time of Aristotle. What is worthy of remark here is the criterion on which Columbus sees the brain as supreme. It is related directly to his view of the three rivers of life in the body. For what is generated in the brain and distributed through the nerves – that is, the animal spirits – is precisely what differentiates the live body from a dead one: 'without the use of the brain man would be like a statue, deprived of sense and motion'.[37] Columbus gives other reasons for the supremacy of the brain in the body,[38] but this is the most important one, that it is the source of sense and motion, that it turns a statue into a live man.

This view of the brain as supreme, and of the three life-giving rivers in the body is all derived from the fact that Columbus constantly used and advocated vivisection. In promoting vivisection as the anatomist's proper and primary procedure, Columbus was consciously and deliberately modelling his work on that of Herophilus and Erasistratus. Columbus devoted a whole book of his *On anatomizing* to promoting the value of vivisection: Book XIV is called 'On vivisection'. This is how it opens – with an attack on Galen and, by implication, on Vesalius:

> While we are zealously pursuing the knowledge of Anatomy with might and main (as they say), several matters sometimes crop up which delay us in our advance, whose action we can in no way know unless we use vivisection. Not indeed the vivisection of men, as the Ancients employed, which to a Christian physician cannot but seem sinful and impious, but of brutes. Particularly since one should notice with regard to the live anatomy of the human body which the pristine physicians employed, that it came to be in such bad repute that not only did it become unlawful to dissect living men, but even to dissect the cadaver of man. And I think that it was due to this that Galen was persuaded that he should abstain from the Anatomy of the human body. But, dear Galen, if it seemed too cruel to you to cut up a living man, if your mind recoiled, if you were in dread; or if you were not permitted to cut up even a dead man, either through the edicts of Rulers or through ancient

> custom, then how was it permissible to you – dissecting apes – to contradict the Ancients, given that you yourself are the most reliable witness that they dissected human bodies? Today indeed, since Popes, Kings and Emperors for the sake of the common benefit of the living allow Anatomists to teach Anatomy in the public Academies annually by dissecting the corpses of criminals, it is easy to solve those issues that you, Galen, disputed. You criticise the Ancients in many places, although you yourself deserve greater criticism than they do! For even if the ape has some similarity to man, yet it is an ape, not a man; nor does its structure (*compago*) correspond in every respect to the fabric of man; and you will see in man several parts about which the ancient Anatomists spoke, which the ape lacks. But we will come back to this.
>
> Since we are prohibited by the Christian Religion from dissecting living men, if there is not an adequate supply of apes or bears or lions, to which the fabric of man comes close – whose vivisection, indeed, whether they were wild or tame, would be very difficult when they were provoked – we shall speak about the vivisection of the dog.
>
> Therefore I choose a young dog (a male or a female as you prefer); for when they are cut puppies yell more than old dogs; it should have a long neck. I suggest dissecting a living dog rather than a pig, because pigs are less suitable for distinguishing the use of the *nervi reversi* [recurrent laryngeal nerves], both because of their excess of flesh, and also because of their horrible-sounding squealing and their grunting . . .[39]

Anatomical problems about actions can be resolved if one uses vivisection, claims Columbus, and the use of vivisection reveals that Galen and Vesalius, and also Aristotle, were wrong on a number of issues. Only systematic vivisection, the method of the ancient Anatomists whom Galen criticized, leads us to the truth.

Of course other anatomists had practised the vivisection of animals. Galen himself and Vesalius are among them. And while the pig had been the human-substitute of preference for most such anatomists, Vesalius had used dogs extensively, as his Bologna demonstrations reveal. So this is not new. Is Columbus's advocacy of vivisection just rhetoric then? No. For what is new is that Columbus is setting out systematically to adopt the ancient Alexandrian approach: the *live* animal is the object of choice for anatomizing. This Alexandrian approach Galen had been opposed to, not because it was vivisectional (for he may not have even known this),[40] but because, as represented by the opinions of Erasistratus and his later followers, it had promoted an attitude and produced results which Galen believed were both absurd and dangerous. So while Galen had not of course been hostile to the vivisection of animals, and had practised it publicly himself, yet, as we saw in an earlier chapter, his grounds for doing so were primarily to attack the views of earlier vivisectional anatomists, particularly Erasistratus! Galen's use of vivisection, then, had been heuristic: it was supposed to offer crowning evidence of the correctness of Galen's views, to convince others to become Galenists. Vesalius had followed Galen in this. But some of the issues Galen thus asserted to be settled were in fact still dubious, either because in the very act of attacking other views (such as those of Erasistratus) he had revealed that his own account was not the only possible account, or because what he himself said seemed to later anatomists to differ from what Aristotle or some other anatomist had said, or (as in the case of Vesalius) because what Galen said was the case did not always seem to conform to experience when tried. It is these issues which Columbus was seeking to resolve by vivisection. That is to say, he was not trying to replicate what Galen did (which had been Vesalius's prime motivation in his vivisectional demonstrations), nor was he trying simply to contradict Galen, nor even trying to prove the ancient Anatomists right. Thus some of his answers confirm that Galen was right, some that that the ancient Anatomists were right, while some show that no one hitherto had been right.

The actions Columbus explores by vivisection include the operation of the voice (that is, by squeezing and cutting of the recurrent laryngeal nerves); the motion of the lung; the motion of the heart, how it is alternately enlarged and constricted; the motion of the arteries, and whether it is the same or opposite in time to the motion of the heart; which motion of the heart is constriction, and which is relaxation; the dilation and contraction of the brain; whether the artery-which-looks-like-a-vein contains air and vapour or blood; and the different kinds of pulse. He comes up with a new view of which state of the heart is constriction and which expansion, and this contradicts the view of Galen. And he concludes that vivisection shows that the arteries pulsate alternately with the heart; that is to say, he concludes, with Erasistratus but against Galen, that the arteries pulsate by the impulsion of blood into them. And he also concludes (and this is against both Galen's and what he took to be Vesalius's positions), that the artery-which-looks-like-a-vein contains blood and not just air to cool the heart. Taken in association with his view of the 'new use of the lungs', this means that Columbus believes that this is blood mixed with *pneuma* – and this position is also against that of Galen, and similar to that of Erasistratus.

Such implications of Columbus's claims would have had to be worked out by his reader. Here in the text Columbus criticized only Aristotle explicitly, not even mentioning Herophilus and Erasistratus by name, though they are obviously the 'pristine physicians' and 'ancient Anatomists' he refers to. The well-informed reader would have recognized the novelty of Columbus's position on these issues, would have seen that he was siding so often with Erasistratus and against Galen. In the light of this we can see why it was that Vesalius had not mentioned the name of

Erasistratus among those who practised anatomy 'in Alexandria in the age of Herophilus, Andreas, Marinus and other most outstanding leaders of dissection' (Chapter 4 above), for to follow Erasistratus was to betray almost everything that Galen had stood for, and that had in no way been Vesalius's intention. But siding with Erasistratus against Galen was something which Columbus did not shrink from doing.

We can conclude about Columbus therefore, that the sequence of events for him must have been first a decision to follow (as Vesalius recommended) the practice of the pristine physicians, the true Ancients, and therefore anatomize the human and vivisect the nearest comparable creatures. This concentration on the body in action led him to adopt a view of the primacy of actions – and hence to treat the source or fountain-head of the actions, the brain, as the supreme organ. By stressing actions he also came to adopt a view of the body as being sustained in life by three rivers flowing from the three major organs. This led him to make a conceptual separation between the operation of the lung and the heart, and this in turn brought him to identify the lung as the place where the vital spirit is first generated, and hence to perceive the pulmonary transit of the blood. Finally, in his anatomical practice Columbus was able to test against vivisection all the doubtful or controverted claims about the operations of the body made by Aristotle, Galen and Vesalius. The true gospel for Columbus was not Galen, as it had been for Sylvius, but the pristine Anatomists of ancient Alexandria. As he wrote, he was outraged 'that some men should swear an oath on Galen's anatomical teachings, so that they dare to affirm that Galen is to be taken as Gospel, and that nothing in his writings is not true!'[41]

(iv)

The first person to announce to the world the new use of the lungs was not in fact Columbus, but Valverde (Juan Valverde de Hamusco, *c.* 1525–*c.* 1588). Valverde came from Spain, and was one of several Spaniards in Italy pursuing medicine and anatomy.[42] He came to be physician to Cardinal Juan Alvarez de Toledo, who was Archbishop of Santiago as well as being the first General Inquisitor of Rome. He dedicated his works to the pope (Paul IV), to his Cardinal patron, and to Philip II of Spain. Like Columbus, Valverde was to be selected out for particular attack by Vesalius, as someone 'who never put his hand to a dissection and is ignorant of medicine as well as of the primary disciplines'.[43] Valverde was a pupil of Columbus in anatomy, possibly beginning in Padua (*c.* 1541), and certainly working with him in Pisa (in 1544), and possibly continuing again in Rome.

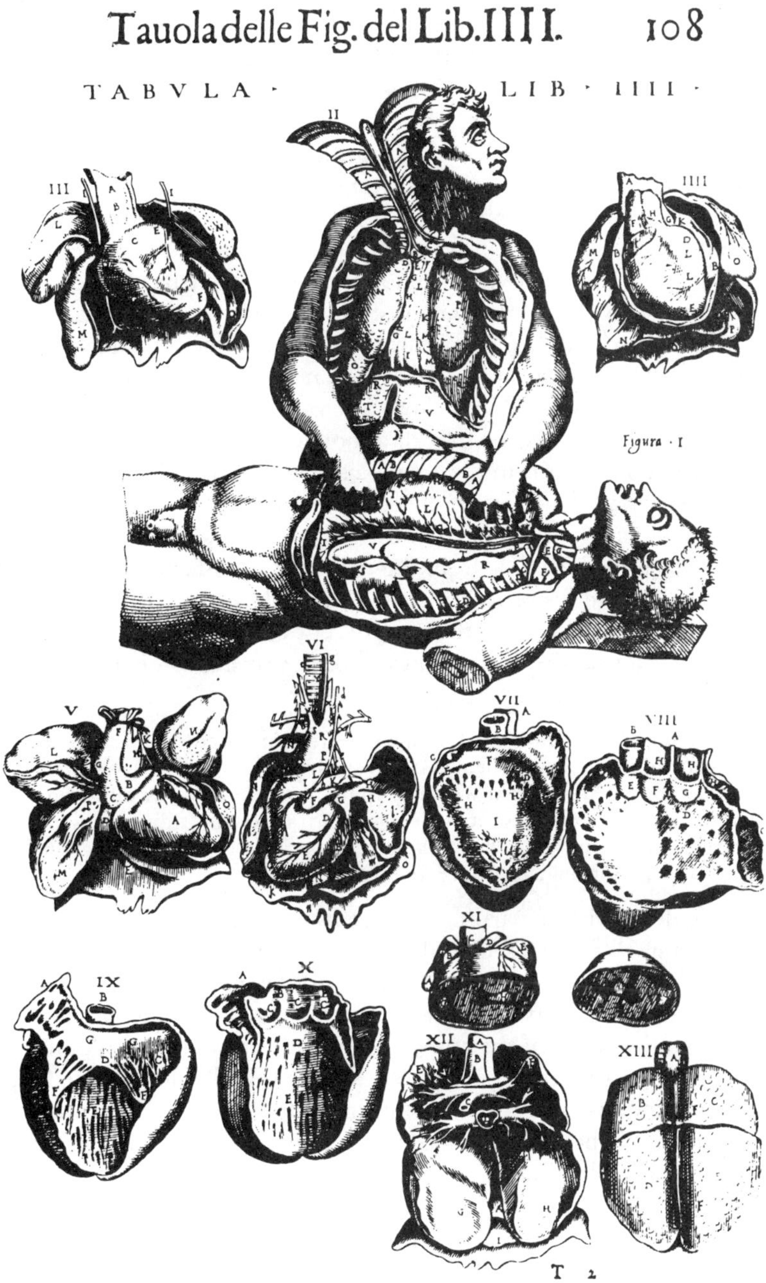

5.4 Page of anatomical figures, copied or modified from Vesalius's *Fabrica*, and printed in Juan Valverde de Humasco, *Anatome Corporis Humani* ..., Venice, 1589, p. 108.

Though he may have written it initially in Latin (for he had a papal licence to print a 'History of the fabric of the human body, in Latin and Spanish'),[44] Valverde published his anatomy book first in Spanish in Rome in 1556, and this was issued in an Italian version at Rome in 1560, printed by the same Venetian printer who produced Columbus's book. Finally it appeared in Latin, translated into that language by one of Columbus's sons, Michael, in 1589. Valverde illustrated his book with illustrations copied or modified from those of Vesalius (see Figure 5.4).[45] In this work Valverde announces that the artery-which-looks-like-a-vein is full of blood, not just spirit, and that the blood passes from the right ventricle of the heart to the left via the lung. He urges his reader to try the experiment, and says that he has performed this many times with Realdus Columbus 'my preceptor'.[46]

But although he was the first to announce it, it is most unlikely that Valverde had much role in the actual discovery of this phenomenon. For it is evident that Valverde's procedures in anatomical investigation are not the same as those of Columbus; Valverde does not follow out the route of each of the rivers from the three fountain-heads of liver, heart and brain. Instead his procedure is to deal with the bones, then with the ligaments and muscles, and then to dissect the three venters of the body, paying only lip-service to the concept of the rivers, as his mixed metaphor reveals:

> There are three recesses, abdomen, thorax and head. In these three seats there are three principal members living, in which as in an oven (*in fornace*) are formed whatever things are necessary for nutrition, life and motion. Now it is necessary to learn that three kinds of vessel (*meatus*) arise from these three principles, imparting their force as if through aqueducts to all the other parts of the body.[47]

Valverde deals with the arteries and veins together in one book as 'they are very alike',[48] and then deals with the nerves in a separate book. Only in this respect is Valverde's sequence different from that of Vesalius. And indeed, as he says, his objective was to offer (like Vesalius) an account of the fabric of the human body, rather than (like Columbus) to write about anatomizing proper. So, although Valverde was the first to announce the discovery of the transit of the lungs, he was not a full practitioner of the new method and a follower of the ancient Anatomists, as was Columbus.

Notes

1. See for example the account given by Gweneth Whitteridge, *William Harvey and the Circulation of the Blood*, London, Macdonald, 1971, pp. 41–77.

2. The major modern sources are Edward D. Coppola, 'The Discovery of the Pulmonary Circulation: a New Approach', *Bulletin of the History of Medicine*, 1957, **31**, pp. 44–77; and Jerome J. Bylebyl's entry 'Colombo' in the *Dictionary of Scientific Biography*, both of which I have used here.
3. See *Realdi Colombi Cremonensis, In almo Gymnasio Romano Anatomici celeberrimi, De Re Anatomica Libri XV*, Venice, Bevilacqua, 1559, p. 60; Coppola, op.cit., pp. 50–1. For a translation of the pertinent passage (where Columbus is referred to, but not directly named) see Jerome J. Bylebyl, 'Cardiovascular Physiology in the Sixteenth and Early Seventeenth Centuries', Unpublished Yale University Ph.D. Thesis, 1969, pp. 164–8, see esp. 167.
4. The resolution about anatomy teaching is printed in Latin in Angelo Fabroni, *Historia Academiae Pisanae*, 3 vols, Pisa, 1791–95, vol. 2, pp. 73–74; it is translated in part in Coppola, op.cit., p. 52, but Coppola is not correct in identifying Columbus as being one of the 'Anatomistae', for this term refers to the students who were in charge of arranging the demonstration (see above, Chapter 5). On Pisa see also C.B. Schmitt, 'The University of Pisa in the Renaissance', *History of Education*, 1974, **3**, pp. 3–17.
5. Fabroni op.cit., vol. 2, p. 470.
6. See Schmitt, op.cit., p. 8.
7. On the early history of the Sapienza see D.S. Chambers, '*Studium Urbis* and *Gabella Studii*: The University of Rome in the Fifteenth Century', in Cecil H. Clough (ed.), *Cultural Aspects of the Italian Renaissance: Essays in Honour of Paul Oskar Kristeller*, Manchester University Press, 1976, pp. 68–110. Otherwise the basic work on the Sapienza is still F.M. Renazzi, *Storia dell'Universita degli Studi di Roma, Detta Communemente La Sapienza*, Rome, Stamperia Pagliarini, 4 vols, 1803; Columbus is discussed in vol. 3, pp. 193–94. Ludwig Pastor, *The History of the Popes*, ed. R.F. Kerr, vol. 8, London, Kegan, Paul, Trench, Trubner and Co., 1908, pp. 271–77, gives an unsympathetic account of the founding of the university.
8. The letter from Columbus is reproduced in facsimile in Coppola op.cit., who offers the above translation.
9. Ibid.
10. See Ascanio Condivi, *The Life of Michelangelo*, translated by Alice Sedgwick Wohl, ed. by Hellmut Wohl, Louisiana State University Press, 1976, p. 99; Giorgio Vasari, *Life of Michelangelo Buonarroti*, translated by George Bull, London, Folio Society, 1971 (from the 1568 version), p. 134 also records this friendship and calls Columbus the 'dear friend' of Michelangelo.
11. This was donated to the Ashmolean Museum, Oxford, in 1846 as a genuine Michelangelo sketch and long thought to be authentic, but has subsequently been attributed to Polidoro and then to Passarotti, and is currently credited to Capitelli. See Ludwig Choulant, *History and Bibliography of Anatomic Illustration*, first published 1852, translated and annotated by Mortimer Frank, New York, Hafner, 1945, repr. 1962, pp. 106–08; K.T. Parker, *Catalogue of the Collection of Drawings in the Ashmolean Museum. Vol. II: Italian Schools*, Oxford, Clarendon Press, 1956, p. 239.
12. Condivi, op.cit., p. 17. On Michelangelo's interest in anatomy in general,

see David Summers, *Michelangelo and the Language of Art*, Princeton University Press, 1981, esp. pp. 20–22, 397–405, 426–29. Valuable also is R.J. Clements's article on Michelangelo in *The Encyclopaedia Britannica*.

13. On the influence of Michelangelo's image of the flayed skin, see Lola L. Szladits, 'The Influence of Michelangelo on Some Anatomical Illustrations' *Journal of the History of Medicine*, 1954, 9, pp. 420–27.
14. Ibid., pp. 92–93; for a comparable account from Vasari, see Vasari op.cit., pp. 120–21.
15. Condivi, op.cit., pp. 98–99. The work of Dürer referred to here is his *Vier Bucher von menschlicher Proportion*, Nurenberg, 1528; facsimile published in 1970 by G.M. Wagner of London. This work also appeared in French and Latin versions. Dürer's approach does not involve any knowledge of (internal) anatomy.
16. For this usage see, for instance, the opening words of the second paragraph of the Proemium to William Harvey's *Exercitatio Anatomica De Motu Cordis et Sanguinis in Animalibus*, Frankfurt, 1628: 'Almost all Anatomists, Physicians and Philosophers to this day suppose, with Galen, that the use of the pulse is the same as that of respiration . . . '; my translation. Columbus does of course also use the term 'anatomicus' in a wider sense.
17. 'Scripserat Realdus Columbus Cremonensis pater superioribus annis de re Anatomica libros XV distincte sane atque accurate: quos cum a doctissimis viris probari, atque expeti intelligeret propediem aediturus erat. Iamque aeneis caracteribus Venetiis describebantur, cum illum nobis praepropera mors invidit'; Dedication.
18. Robert J. Moes and C.D. O'Malley, 'Realdo Colombo: "On Those Things Rarely Found in Anatomy"', *Bulletin of the History of Medicine*, 1960, 34, pp. 508–28, p. 509 (following Renazzi, op.cit., vol. 2, p. 193), claim that the title-page is by Paul Veronese, but offer no supporting evidence.
19. Moes and O'Malley, op.cit., p. 511 (following Renazzi, op.cit., vol. 2, p. 193) say that copies of *De Re Anatomica* exist which carry Columbus's own dedication to the previous pope, Paul IV (Carafa); I have not been able to see any such copy.
20. 'Cum mihi in mentem venit . . . de universa Anatome primum, deinde vero de anatomica administratione, quae observavi, et cum rei natura consentire experimento didici, scribere'.
21. Candido Lectori; my translation.
22. On Servetus see Part II, Chapter 8 below.
23. The viscera which Columbus deals with in Book XI, are: the soft palate, tonsils, the lining of the nose, the palate, mouth, oesophagus, stomach, intestines, and trachea ('hard artery'); the lung; the pleura and mediastinum; the mesentery and omentum; the spleen; the gall bladder; kidneys; bladder; peritoneum; abdomen; seminal vessels; testicles; penis; uterus; umbilicus. On the eye Columbus says that it is 'a most noble part of our body' (p. 123); it is, for Columbus, the direct connection with the brain that makes the eye so noble, and so he devotes a whole book, Book X, to the eyes.
24. I use the expression 'Latin anatomizing' here to separate what I am talking about here from discussion of Ibn al Nafis and whether he 'deserves the credit' for the discovery of the pulmonary transit of the blood.

The matter of priority of discovery is not my concern in this book, but on it see for instance L.G. Wilson, 'The Problem of the Discovery of the Pulmonary Circulation', *Journal of the History of Medicine*, 1962, **17**, pp. 236–44. For the translated text of the pertinent part of Ibn Nafis's commentary on Avicenna, see E. Edward Bitter, 'A study of Ibn Nafis', *Bulletin of the History of Medicine*, 1955, **29**, pp. 352–68 and 429–47.

25. In the chapter 'On the transverse septum', which Columbus discusses in the section on muscles, all he says is this: 'The diaphragm is the muscle dividing the natural members [vitalia, but corrected in Errata] from the vital, as Divine Plato says, who used to divide the irrascible from the concupiscible soul'. Columbus does not directly say whether he agrees with this or not, but he immediately goes on to reject the view of Aristotle (*sic*) 'that this division arose from nature' in order to keep the vapours of the food away from the heart. *De Re Anatomica*, p. 140, my translation.
26. It is from this that we can infer that the sequence of the text represents Columbus's approach to anatomizing, and is not just a convenient means of exposition adopted later for the purposes of the book. Columbus also comments on several occasions on the unusualness of his order, as over the muscles of the hand (pp. 156–57), which he dissects last of all the muscles (rather than the first, as Galen and Vesalius did), and over the larynx, which he treats among the bones (p. 43).
27. *De Re Anatomica*, p. 51.
28. Ibid., p. 43.
29. Ibid., p. 55.
30. Ibid., p. 224.
31. Ibid., pp. 183, 189 (this accords with the view of Galen), 191, respectively.
32. Ibid., pp. 220–21, my translation.
33. I am indebted to Professor Vivian Nutton, whose knowledge of Galen is probably unrivalled, for assuring me that Galen does not use this image of 'rivers' flowing from a fountain-head. For Galen specifically saying that the heart is *not* the source of the arterial system in the sense in which a river has a source, see David J. Furley and J.S. Wilkie, *Galen on Respiration and the Arteries*, Princeton University Press, 1984, pp. 30–31. For an example, by contrast, of Galen using the image of channels irrigating a garden, see Peter Brain, *Galen on Bloodletting: A Study of the Origins, Development and Validity of his Opinions, With a Translation of the Three Works*, Cambridge University Press, 1986, pp. 154–55. Aristotle had similarly written in *Parts of Animals*, 668 a 14–22 of the blood vessels being like water-courses in gardens; see Furley and Wilkie, op.cit., p. 21. On the whole issue of organs as *archai* (principles, beginning points), see I.M. Lonie, 'Erasistratus, the Erasistrateans, and Aristotle', *Bulletin of the History of Medicine*, 1964, **38**, pp. 426–43.
34. *De Re Anatomica*, p. 185.
35. Ibid., p. 223.
36. Ibid., pp. 164, 175, 188.
37. Ibid., p. 188.
38. Ibid., pp. 187–89.
39. Ibid., pp. 256–57, my translation; Bylebyl, op.cit. p. 176, translates the opening of this passage differently.
40. It should be remembered that Galen never mentions that Herophilus and

Erasistratus performed vivisection (if indeed they did); we know about this, as Columbus did, from Celsus. It may be that Galen never knew that they had supposedly performed vivisection, but he certainly knew that they had dissected humans.

41. 'Quid? quod aliqui nostro tempore in Galeni Placita de anatome iurarunt; ut hoc audeant affirmare Galenum Evangelistae more suspiciendum esse, nihilque in eius scriptis esse non verum', Columbus op.cit., p. 179, as translated by Bylebyl, op.cit., p. 306. A contemporary anatomist, Fallopius, wrote similarly: 'when Galen refutes Herophilus, I consider him to be refuting the Gospel itself', and 'with me the authority of Herophilus on anatomical matters is the Gospel', Gabriel Fallopius, *Observationes Anatomicae*, Venice, 1562, as quoted by Heinrich von Staden, *Herophilus: The Art of Medicine in Early Alexandria*, Cambridge University Press, 1989, p. xi, my translation.
42. The most important work on Valverde has been done by Francisco Guerra, 'Juan de Valverde de Amusco', *Clio Medica* 1967, 2, pp. 339–62, whom I follow here.
43. Ibid., p. 354, quoting Vesalius's *Examen* of the opinions of Fallopius (1564).
44. Ibid., p. 343.
45. The most complete translation imaginable occurred when these illustrations were re-cut for one of the Latin editions since in it, for some reason, the resulting prints are all back to front. As the human body is mostly symmetrical this causes important problems only when the heart is portrayed; and the heart is, duly, portrayed as on the right side of the body! The 1589 Venice edition has the figures the right way round, but they are reversed in the Venice, Junta, edition of 1607.
46. Guerra, pp. 356–57, 355, translating Valverde.
47. My translation from p. 263 of the Latin edition (Venice, 1607).
48. '*Una etiam arterias persequar, quam venis, ut plurimum sociae accedant*', ibid., p. 263.

CHAPTER SIX

Fabricius: The Revival of Aristotelian Anatomy

We have seen so far a progressive recovery of the anatomical practices of the Ancients:

Aristotle	Herophilus/Erasistratus	Galen		Mundinus
c. 350 BC	*c.* 250 BC	*c.* 200 AD		*c.* 1316 AD
	↑	↗	↖	↑
	Columbus	Vesalius	Sylvius/Guenther	Berengario
	c. 1550	1543	1530	*c.* 1520

As the table shows, Berengarius in the first two decades of the sixteenth century tried to recover the true text and meaning of Mundinus of *c.* 1316; then Sylvius and Guinther in the 1530s tried to recover the true *text* of Galen of about 200 AD; then Vesalius in the early 1540s tried to recover the *practice* of Galen, and then rejected it in favour of a more ancient practice; then Columbus in the 1550s tried to recover the practice of the early Alexandrians of the third century BC. From taking Galen as the exemplar (as Vesalius did) to taking the early Alexandrians as the model for anatomizing (as Columbus did), involves a move from concentrating on the *fabric* of the human body to concentrating on its *actions*. But no one has yet shown an equivalent obsession with the *uses* or *causes* of the parts of the body.

One Ancient, in many respects the most important of all, and for anatomy easily the most ancient, has not been dealt with in this story of the recovery of ancient practices, because we have not yet found anyone trying to resuscitate his practice of anatomy. That Ancient is of course Aristotle. We have seen that the practices of the Ancients were recovered in reverse chronological sequence, so we might expect the practice of the ancientest anatomical Ancient to have been recovered last. This was indeed the case. Aristotle's anatomical practice was to be reconstructed and followed anew only in the second half of the sixteenth century. Of course there is one sense in which it is true that all anatomists after Aristotle had indeed been following Aristotle. For every subsequent anatomist does indeed quote some of Aristotle's doctrines, deploy some of his concepts, argue with one or other of his claims. What is more, Galen's anatomical project, on which (as we have seen) so much later anatomical activity was in turn built, would have been inconceivable without its base

– its acknowledged base – in the work of Aristotle. But following Aristotle, quoting his doctrines, deploying his concepts, arguing with his findings, even being obsessed with his thinking, is not the same thing as practising Aristotle's *own* project as Aristotle himself would have done. For, as we saw earlier, Aristotle thought of, demarcated and investigated the body in a way quite other than any we have hitherto met with among the anatomists who came after him; and most remarkable of all, of course, is the fact that Aristotle's object of anatomical enquiry had not been man at all! Thus it is clear that no anatomist at all after Aristotle had actually pursued the *same* project of anatomical enquiry as Aristotle himself. They had integrated Aristotle's doctrines, concepts, claims and information into their own *differing* projects.

The sixteenth-century return to Aristotle the anatomist was of a piece with a particular, larger programme of return to Aristotle the whole philosopher, a programme which produced major novelties in the understanding of what Aristotle was saying and what philosophy was about. Aristotle had of course been the mainstay of the curriculum since the thirteenth century when universities began, with pretty much the same selection of his books being read and commented on over the intervening centuries. Modern studies are revealing that the sixteenth century saw perhaps the greatest flourishing of the study of Aristotle – whereas historians had hitherto assumed that Aristotle's texts were dropped in the sixteenth century in favour of anti- or non-Aristotelian approaches.[1] Every educated man knew his Aristotle.

The particular revival of Aristotle with which we are here concerned, which involved a general programme of resuscitating the works of Aristotle *together with* the revival of his anatomical project, was unique to Padua, the home university of the republic of Venice, and it was promoted by the committee of *Riformatori* who ran the university for the Venetian senate. We may start our view of teaching at Padua by casting our eyes over one of the printed announcements of lectures, taking for convenience the one for the session beginning November 1593 as our example.[2] There are 26 professorships listed, including one vacancy. On the original document, in addition to the names of the professors, the hours of lecturing and the texts are also specified for each position. 'Ordinary' professors are senior to 'extra-ordinary' ones; the chairs 'in the way of St Thomas' (St Thomas Aquinas) and 'in the way of Scotus' (Duns Scotus) were usually occupied by Dominicans and Franciscans respectively. The list is in the published order and indicates the hierarchy of the chairs and subjects:

Chair in Theology in the way of St Thomas
Chair in Theology in the way of Scotus

Chair in Holy Scripture
Chair in Metaphysics in the way of St Thomas } *Metaphysics* Book 1
Chair in Metaphysics in the way of Scotus } *Metaphysics* Book 1
Two Ordinary Chairs in the Theory of Medicine
Two Ordinary Chairs in the Practice of Medicine
Two Ordinary Chairs of Philosophy } *On the soul* Books 1 & 2
Two Extraordinary Chairs in the Theory of Medicine
Two Extraordinary Chairs in the Practice of Medicine
Two Extraordinary Chairs of Philosophy } *On the soul* Book 3
Chair in the Moral Philosophy of Aristotle (vacant)
Chair in Surgery and Anatomy
Chair in the Third Book of Avicenna
Chair in Simples [i.e. plants for medicines]
Three Chairs in Logic } *Posterior Analytics* Books 1–3
Chair in Mathematics
Chair in Greek and Latin *Humanitas*

To a modern eye, the amount of Aristotle being read is quite astonishing: the two professors of metaphysics both read the same work by Aristotle, the *Metaphysics*; the two ordinary and the two extraordinary professors of philosophy all read Aristotle's *On the soul*; there is a professor of the moral philosophy of Aristotle, who would normally read Aristotle's *Ethics*; and the three professors of logic read the first three books of Aristotle's *Posterior Analytics*. This teaching list of course reveals only part of the complete cycle of texts to be read by the professors. For instance the metaphysics professors had to read not only book 1 but also books 7 and 12 of the *Metaphysics*. Similarly the ordinary and extraordinary professors of philosophy had to read books 1, 2 and 8 of Aristotle's *Physics*, the two books of *On generation and corruption*, the three books of *On the soul*, and the four *On the heaven and the earth* in the course of the full cycle of lectures, while only their reading of *On the soul* is reflected here in this list.[3]

While this is fundamentally still a fourteenth-century curriculum being taught in the late sixteenth century, it had received some modifications in the interim, though the specified texts of Aristotle had not been changed. The two chairs of theology and the two of metaphysics were a response (common to many universities) to late fifteenth-century demands for rival interpretations by the rival orders of friar; the chair in Holy Scripture had originally been a chair in the *Sentences* of Peter Lombard, and was in 1551 updated to be on the texts of the Old and New Testaments. The chair in mathematics (held in 1593 by Galileo) had been established in 1517 to teach 'astronomy and mathematics', in other words astrology for medicine. At Padua most of the professorships had been established in pairs, so that each had an antagonist or *concurrens* professor of different views who lectured at the same hours and on the same text; this is the case wherever in the list there are two chairs of the same subject and status. This unusual system promoted a great deal of healthy controversy between colleagues, which often spilled over into print.[4]

We can see that, as had been traditional at Padua, medicine and philosophy (arts) are still studied and taught in one joint faculty as they had been in the time of Mundinus, and that Paduan medical students were expected to study philosophy with their medicine; that through the system of concurrent professorships this philosophy was deliberately multifarious in interpretation but that it exclusively discussed Aristotle and the issues raised by his writings; that it was the philosophy professors who taught the 'nature' books of Aristotle; and that Aristotle's book on the soul, *De Anima*, played a major role in the philosophy teaching of those same professors. To put it another way, the question of the soul – Aristotle's view of the soul – was basic to the teaching of Natural Philosophy, and that a renewed Aristotelian Natural Philosophy was, in turn, fundamental at Padua to both medicine and anatomy. To the significance of this we now turn.

The Paduan philosophers were concerned with trying to reach Aristotle's own sentiment. One of the most famous of them, for instance, giving an oration to the students 'In praise of the study of philosophy' at Padua in 1585, said 'I will arrange what I am going to say according to two heads: for since our aim is to be skilled in the philosophy of Aristotle, we must see in the first place what mode of philosophising Aristotle himself used, and then find what we need to do so that we may be said to philosophise correctly and "aristotelically" on Aristotle.'[5]

Randall has traced the way in which discussion of the soul, of its capacity to acquire knowledge and of the mode of its existence, was conducted at Padua in progressively more 'aristotelical' a mode in the course of the sixteenth century, and he has indicated how it came to be divested of Platonic and Averroist interpretations by giving privileged attention to the ancient Greek commentary of Alexander of Aphrodisias.[6] Beginning with Petrus Pomponazzi in the early decades of the century, this culminated in the period 1564–89, when Jacobus Zabarella was one of the professors of philosophy. In Zabarella, Randall claims, 'Aristotle's thought has been made so transparently his own that it seems unjust to call him a follower: it is Aristotle himself, speaking in the Latin of Padua – not the syllogistic Aristotle with a category for every emergency – but the Aristotle who insists on the primacy of subject-matter, of fact, of experience.'[7] Zabarella's time as Paduan professor coincided with that of Caesar Cremonini (professor 1573–1631), and Franciscus Piccolomini (professor 1560–1607), who also had reputations as being new Aristotles – it being said of Cremonini, for instance, that he was 'the Genius of Aristotle'[8] – though they did not necessarily agree with each other in every detail of their understanding of Aristotle.

In general these Paduan Aristotelian philosophers claimed that the soul is the act of the body. As Zabarella put it:

> The body is not the act of the soul, but the soul is the act of the body; it is not body, but something belonging to body, namely its act and perfection, whence the soul cannot exist without the body, since a perfection cannot exist without that of which it is the perfection . . . He who does not see that Aristotle plainly says that every soul is an informing form, is blind.[9]

If the soul cannot exist without the body, then the implication is clear: that when the body dies the soul dies; or, rather, the death of the body *is* the death of the soul. The soul is mortal. And this is contrary to Roman Catholic and Protestant teaching about the immortality of the soul. And if the soul indeed cannot exist without the body, then a second implication is also clear: that the soul can be understood *only* through its operations in the body. If we want to understand what makes a man a man, or a plant a plant – that is, their soul, their informing form – then we can only do so by studying the operations of their respective bodies, for the soul 'is the act of the body'.

(ii)

The man who resurrected Aristotle's research programme in anatomy was the Paduan Professor of Anatomy and Surgery, Hieronymus Fabricius ab Aquapendente (Girolamo Fabrici or Fabrizie), whom we shall be referring to by his Latinized name as Fabricius.[10] It was Fabricius (see Figure 6.1) who introduced this new Aristotelian anatomy at Padua, and who first read and practised Aristotle 'aristotelically' in anatomy. He held the post of Lecturer on Surgery and Professor of Anatomy from 1565, when he was about 32 years old, until 1613, when he was about 80, the last dozen or so years with the title 'Professor Superordinarius'. His winter months were dedicated to anatomy, his summers to surgery.[11] It was during Fabricius's reign as Professor of Anatomy that the fabulous Paduan anatomical theatre was built (completed 1594) in the new university buildings (see Figure 3.5, Chapter 3 above).[12] Fabricius is customarily credited with first discovering the so-called 'valves in the veins', though he did not think of their function as valves (as we shall later see).

6.1 Fabricius of Aquapendente.

Though he held his chair for so many years, Fabricius had an uneasy relationship with the students who came to Padua to learn anatomy. Many objections were made against him by generations of students for failing to perform his duties and for teaching the wrong things.[13] Yet Fabricius was nevertheless maintained and promoted in his post by the *Riformatori*, who kept increasing his salary. As the presence of the students was what made the university a successful business for Venice, the continued support of Fabricius by the *Riformatori* indicates very strong commitment on their part to the content and nature of his teaching and their conviction that what he taught was what the students really needed, even if some of them occasionally did not appreciate it.

That teaching, from Fabricius's earliest years in post, was Aristotelian. Reading Aristotle 'aristotelically' on anatomy, Fabricius of course read from *De Anima* ('On the soul') outwards, and hence his anatomizing dealt with the different faculties of the soul and his object of inquiry was 'The Animal'. Like Aristotle, he looked at the body as the soul-in-action; or, in the words of his own colleague for many years, Zabarella: 'the soul is the act of the body'.[14] Fabricius's anatomical work was thus primarily about the soul, and about the body itself only in a secondary sense. Thus what Fabricius was studying was the soul, and he was teaching it to students who had already had the opportunity of acquiring extensive exposure to Paduan ways of understanding Aristotle on the soul.

Some of Fabricius's research and teaching is recorded, in a refined and revised form, in his one great written work. This is a great book which he set about publishing in his old age, made up of part-numbers, many but not all of which he succeeded in finishing and publishing before his death, and it was called *Theatrum totius animalis fabricae*, 'A Theatre of the Whole Animal Fabric'.[15] The theme, it will be noted right from the title, is not the Galenic or Alexandrian one of 'man' but the Aristotelian one of 'The Animal', and Fabricius's usage of the term 'fabric' was much wider than that of Vesalius. As in the case of Aristotle himself this does not mean that Fabricius was engaged in 'comparative anatomy', nor in 'biology', but in that now-lost study, the study of 'The Animal' and its 'oeconomy', a discipline which for Fabricius was part of Philosophy. Man was treated by Fabricius as one kind of animal, but only one. These are the works which went toward making up *The Theatre of the Whole Animal Fabric*:

1. On vision, or on the eye the organ of sight, 1600
2. On the voice, or on the larynx the organ of the voice, 1600
3. On hearing, or on the ear the organ of hearing, 1600
4. On the formed fetus, 1600
5. On speech and its instruments, 1603
6. On the vocal communication of brutes, 1603
7. On the *ostiola* in the veins, 1603
8. On the artifice of the muscle, 1614
9. On the articulations of the bones, 1614
10. On respiration and its instruments, 1615
11. On the local motion of animals as a whole, 1618
12. On the gullet, stomach, intestines, 1618
13. On the covering of the whole animal [i.e. on the skin, etc], 1618
14. On the formation of the egg and of the chick, 1621 (posth.)
15. On the semen (planned but not completed)

To illustrate the extent to which Fabricius's project is identified with that of Aristotle, we may tabulate the contents of his *Theatre* against the soul anatomy of Aristotle (see Chapter 1, pp. 18–19):

Aristotelian aspects of soul	**Works by Fabricius (parts of the *Theatre*)**
VEGETATIVE SOUL:	On respiration 10.
	On the generation of animals:
	On the semen 15.
	On the formation of the egg and of the chick 14.
	On the formed fetus 4.
	On the *ostiola* in the veins 7.
	On the gullet, stomach, intestines 12.
SENSITIVE SOUL:	On vision 1.
	On hearing 3.
	On the covering of the whole animal 13.
MOTILE SOUL:	On the artifice of muscle 8.
	On the articulations of the bones 9.
	On the local motion of animals 11.
RATIONAL SOUL:	On the voice 2.
	On speech and its instruments 5.
	On the vocal communication of brutes 6.

The coherence of Fabricius's works in and with the Aristotelian scheme of anatomical things is clear from this Table. Such topics had simply not been pursued by previous anatomists in this way, since they are Aristotelian through and through. Fabricius's commitment to this specifically Aristotelian project in anatomy is not something that he justifies at any length in his published writings, but after being engaged in successfully teaching and researching on it for some 35 years before publishing, and having outlived his rivals, he may no longer have felt any need to.[16] Fabricius regarded his task as that of going further in the direction indicated by Aristotle: he did not think that Aristotle had said it all, and that his own role should consist only in teaching the established truths of Aristotle. His colleague Zabarella expressed this sentiment thus:

> Although I wish to be second to no mortal in admiration of the great genius of Aristotle, still I believe that he neither could have written everything or known everything, nor did he so follow the truth in all that he did write that he was never able to err; for he was a man, and not God.[17]

And here is Fabricius himself talking similarly about how he, in anatomy, is equally dedicated to Aristotle, and yet is also intent on continuing the anatomical inquiry beyond Aristotle. He is writing about *the generation of animals*:

> few of the Ancients and none at all of the Moderns have addressed themselves to this topic. Why this has happened I do not know, since indeed it is unworthy that such great marvels of nature should lie hidden from us. We shall reveal them with as much brevity as we can, and so effect it both by the placing of illustrations and by the plan of exposition, that anyone henceforth may be able by himself to understand and contemplate those first beginnings of the life of every animal. In this way we shall both follow and expound that great interpreter of nature, Aristotle, who first and alone enquired into those mysteries; and if anything at times escaped him, we shall point it out.[18]

Only Aristotle, that great interpreter of Nature, has pursued this topic of the generation of animals before; we shall add to his work 'if anything at times escaped him'. Fabricius made similar claims about the study of the *local motion of animals* and about *respiration* – that he was the first to resume the study of these subjects since Aristotle.

It is evident from the Table above that Fabricius found topics for inquiry by simply following Aristotle; he investigates 'The Animal' as soul-in-action, and especially the vegetative, sensitive and motile aspects of the soul. In practice that means he concentrates on investigating *processes*, such as generation and respiration, and the organs involved in carrying them out, and in *all animals*. This is, as it had been for Aristotle, anatomy as Philosophy. There is no room in this inquiry for the traditional 'three venter' division of the body, nor for the skeleton-muscles-arteries-veins-nerves sequence of Vesalius, nor for the 'three rivers' approach of Columbus, all of which approaches had anyway taken man as their central object of study. Fabricius's new object of study meant new things became visible, because new questions were being asked. It was an open-ended research programme. The capacity of this approach to find anatomical novelty – including, in time, the circulation of the blood – makes it desirable that we give an outline of Fabricius' actual practice of anatomizing here.[19]

A full account of any organ or part involved, for Fabricius, a four-fold presentation:[20]

1. the *historia*, or description of the structure
2. an enquiry into the *action* of the part
3. a specification of the *uses/usefulnesses* of the part
4. a *demonstration* of the truth and applicability of his findings.

Only such an account was properly 'philosophical' by Fabricius's standards. Fabricius was more than once critical of Vesalius for having performed only the first of these, and hence for not having given a properly philosophical account of anatomy. Fabricius's employment of the fourth stage, demonstration, may well owe something to the discussions of his Paduan philosopher colleagues, and their development of the '*regressus*' technique in this period.[21]

To produce one of his philosophical anatomical accounts, therefore, Fabricius went through a particular routine of investigation. We must bear in mind that Fabricius is always investigating some operation of the soul, such as nutrition, generation, sensation, or motion, and then looking at the organs which carry it out. Thus his goal in any particular anatomical investigation was a *single, general, universal account* of the operation in question in that single, general, universal creature 'The Animal', and the organ(s) or part(s) which carry it out, for this would capture the essence of that particular faculty and instruments of the soul, and hence explain what it was and why it was as it was.

(1) First he had to create the *historia*. This involved looking at as many instances as possible of the part in question in as many kinds of animal as he could. Thus if, for instance, he was investigating respiration, and hence looking at the lung, he would investigate the incidence of the lung in all the creatures that have a lung, and also he would note that the lung was absent in certain creatures (such as fish and insects), and note what other organ (if any) seemed to be standing in place of it. This involved asking a regular series of questions every time; these questions were based on the 'categories' of Aristotle. He asked: is it present? how many present? what form or shape does it have? what position is it in? what connections does it have? and so on. (1a) Fabricius then took this data from the series of separate incidences, and looked for the *similarities* in the answers to each of these questions. Hence he could make an 'induction' and reach the *general* incidence. He could now say (for instance) that in the accomplishment of (say) vision, *it is usually the case* – it is the norm – that part x does occur in animals, it is usually of shape y, it is in position z, it is connected to other parts a and b, and so on. This is the general *historia*. When he came to write up this *historia* for publication, Fabricius would now write in a general, universal way, talking for instance of '*the* eye' rather than of the eyes of particular animals. He would thus have reached the general through a study of the particular. (1b) At the stage of assessing the similarities, Fabricius would also need to make a listing and grouping of the *dissimilarities* in the answers to his set of categorical questions, because his final account would need to be able to explain these as well, would need to be able to account for why in some animals there is no eye, why

it is in some animals in an unusual position, why it is structured in an unusual way, and so on. For, unless the general, universal, account allowed one to answer these questions satisfactorily, then it was not a proper general, universal, account. Ultimately he would be able to give answers about these dissimilarities in terms of the particular requirements or peculiarities of the life, activities, habits or other parts of the particular animal. This is just like Aristotle's procedure of course.

Then (2) Fabricius turned to investigating the particular *action* of the part in question. An action in this sense is some unique and public role contributing to and indispensable to the effective functioning of the whole animal, such as seeing, hearing, making blood, or whatever. As a good Aristotelian Fabricius would assume that the capacity or faculty of a particular part to perform a particular action usually resided in the special nature of the material of that part; thus the liver has a blood-making faculty, the heart a pulsatile faculty. So in identifying and localizing the action, Fabricius would be looking for the special and unique kind of material constituting the part, in whose substance the unique action was localized; for instance, he discovered that the crystalline humor of the eye is where the unique action of the eye-seeing – takes place.

In the next place (3) Fabricius would try and discover the *use* or *cause* of the part in question: the general account of why it exists and why it is as it is in The Animal. For instance, of the placenta, the 'first and chief usefulness', Fabricius discovers, is 'the protection and defence' of the uterine vessels.

Finally (4) Fabricius would give a *demonstration* of the truth of his account. Having ascended from the particulars to the general by induction (from the animals to The Animal), such demonstration could be used to descend from the general to the particulars, and thus show how the particulars are comprehended in – are instances or instantiations of – the general. In the case of anatomy such a demonstration shows how the incidence of a given part as it occurs in each and every instance of animal that actually exists in nature accords with, and is an instance of, the *general* account of it in The Animal. But in practice Fabricius tended to use this demonstration only to deal with particular variations from that norm in particular animals. What he therefore usually demonstrates is how the particular incidence (or absence) of the part in certain animals suits, is necessary for, and is explained by their particular form of life, or activities, or habits, or other parts. That is to say, he shows how they are indeed particular instances of the general case, and not exceptions to it.

We can illustrate this way of proceeding by looking at Fabricius's most famous discovery, the *ostiola*, the so-called valves in the veins. Fabricius claimed to have discovered these in the human around the year 1574 (he was, of course, obliged to demonstrate human anatomy to the medical students). What he had found were little membranes which occur only in the veins, and, then again, only in some of the veins. There are no such membranes in the arteries even though the function and structure of the arteries is comparable to that of the veins. So Fabricius has to be able to explain – and his research will not be finished until he can satisfactorily explain – why these membranes exist in some veins but not all veins, why they do not exist in the arteries (why they are not needed), what their action is, what role or 'use' they fulfil when present in the oeconomy of The Animal, and what the conditions are that make them unnecessary when they are absent. We cannot, of course, read Fabricius's published account as if it were a step-by-step account of his research, like a laboratory notebook; it presents only the results of his inquiry, and in a polished form. But if we bear this in mind, a great deal about his inquiry can be seen from simply looking at the skeleton of the argument of his little book. Fabricius is, of course, taking it for granted that the veins are for the distribution of nutritive blood to all parts of the body, and the arteries for the distribution of 'vital' blood.

> *Definition*: What I am referring to by this name '*ostiola*': delicate membranes in the cavity of veins, occurring singly or in pairs, mostly in the limb veins; opening upwards, and having a form like a node in a twig.
> *General purpose or 'use'*: To *delay* the blood, in the interest of the proper distribution and assimilation of nourishment throughout the body.
> The *ostiola* are necessary in order
> (a) to ensure that the 'upper' limbs receive adequate nourishment;
> (b) to prevent permanent swelling in the extreme ends of limbs.
> No anatomist has discovered them hitherto. Why not? Maybe because
> (a) since the veins are intended for the free flow of blood, anatomists would not expect to find membranes in them;
> (b) they do not occur in arteries;
> (c) they do not occur in all veins (for instance the vena cava, the jugulars, the 'outer' veins).
> Why arteries do not require them:
> (a) arteries have a different role: although they contain blood they are not concerned with nutrition;
> (b) arteries have a different structure: their thick walls are unlikely to suffer distension;
> (c) arterial blood has a different movement: in arteries there is constant flux and reflux, rather than predominantly one-way flow.
> Why was it necessary to retard the flow of blood in the veins? In order to ensure appropriate delay for aliment to be assimilated by the parts.
> Why small veins do not need them:

> (a) they contain only a small amount of blood and all that suffices for them for purposes of local nutrition;
> (b) their needs are met by the action of the *ostiola* in the larger veins;
> Another need for the *ostiola* in the limbs (where they mostly occur): the frequent local motion which is characteristic of the limbs creates local heat; without the existence of the *ostiola* this would naturally draw more blood to the limbs, hence creating undernourishment of the principal parts and rupture of the limb veins;

'either of which', Fabricius continues,

> was going to be very pernicious to the whole animal (*toti animali*), given that it was essential that the principal parts such as the liver, heart, lungs and brain should always abound most copiously with blood. It was for this reason, I believe, that the vena cava (where it passes through the trunk of the body) and similarly the jugulars, should be quite destitute of *ostiola*. For it was requisite that the brain, heart, lungs, liver and kidneys – which procure the conservation of the whole animal *(totius animalis*) – should abound with nourishment, and it was essential that it should not be detained in them even for a moment, both in the interests of replacing lost substance, and of producing the vital and animal spirits whereby life is conserved for animals (*animalibus*).
> But if you observe *ostiola* at the beginning of the jugular veins in man, you may say that they have been placed there to detain the blood, so that in the declined position of the head it should not flood into the brain like a river and be accumulated there more than is appropriate.[22]

We can see from this that Fabricius deals with the *ostiola* as phenomena of the whole animal (*totius animalis*), giving a *historia* of them (number, form, site, distance and so on), an account of their action, and of their use. At all points his account is a *general* one. In the last paragraph quoted, for instance, Fabricius gives the general case on the incidence of the *ostiola* in the jugular veins, based on observation – that they are devoid of *ostiola* – and he gives the general reason for it in the life of The Animal; that here the blood must suffer no delay since the brain, as one of the principal organs, needs an unchecked supply of fresh nourishment. He then deals with an apparent exception, the case of man, where *ostiola* may be found in the jugulars. This apparent exception is resolved by relating it to the particular characteristics of the life of this particular animal. The *general* reason for the presence of the *ostiola* (even here) still holds good: to delay the blood. But man usually carries his head upright, and the supply of nourishment to his brain is suitably catered for in that position. So, when man bends his head, a rush of blood would occur to the brain: hence the (exceptional) presence of *ostiola* in the jugulars of man, fulfilling a special need for the life of this particular animal.

Thus even in a treatise clearly first inspired by his having discovered the *ostiola* in a human subject, and a treatise illustrated exclusively by pictures of the *ostiola* in man, Fabricius is still dealing with the incidence of the *ostiola* in all animals – and hence in The Animal – of which man is but one instance. Finally we can see why Fabricius called these membranes *ostiola*, 'little doors' (rather than 'valves', as this term is usually translated nowadays), for this is how he saw their use or function.[23] When the part is properly identified in this way, then the inquiry is well and truly finished, for the part can now be named from its use or function.

Whenever he published his final results of a piece of anatomical research Fabricius was always dealing with *the general*, but he had a choice about the precise form in which to present it to his reader. He could if he chose present the general account with relation to one particular instance, pointing out the ways in which the chosen example itself differed from the norm or, where this one substantially represented the norm, the ways in which certain other animals differed from this norm in the part or process in question. *On the Formation of the Egg and of the Chick* is an instance of his choosing this form of exposition, with the chicken as the example; so too is *On the Little Doors in the Veins*, where he uses man as his example (see Figure 6.2). Alternatively, he could discuss the norm about a given process or part, and use illustrations of whichever animals most clearly or conveniently showed particular points. This is what he does for example in *On the Formed Fetus*, whose illustrations therefore show a wide range of animals (see Figure 6.3). But still, of course, this is not 'comparative anatomy' he is engaged on, but always the anatomy of 'The Animal'.

Fabricius's concern with 'The Animal' rather than with just man, is further shown by the breadth of the range of illustrations that he used. He wanted the *Theatre* to have more than 300 illustrations, with the parts represented in their natural size 'and, what is no less important, in their natural colour', and in the interests of clarity to have each picture in two forms, one coloured, one not.[24] The coloured ones would presumably have been made under Fabricius's direction by people hand-colouring the monochrome prints at the printer's shop. Some of the published works, such as those on generation, have their full complement of (monochrome) illustrations. But some of them were never published. Many of the original coloured paintings that Fabricius prepared are now in the library of his homeland, St Mark's Library in Venice.[25]

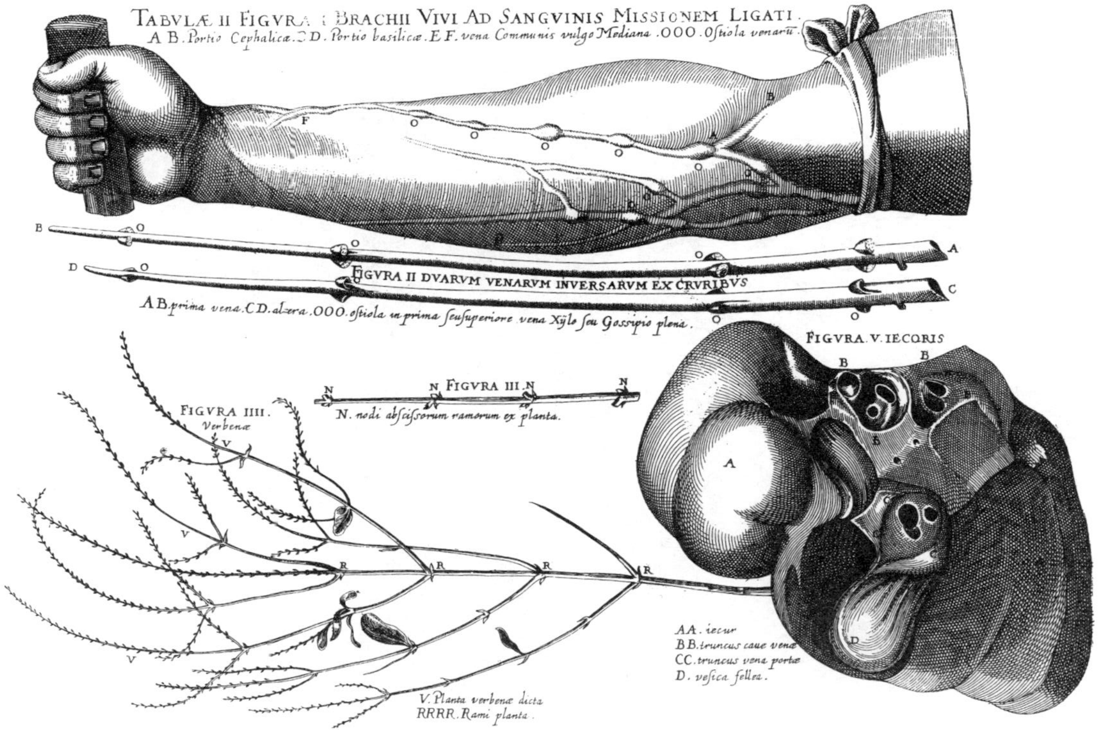

6.2 Fabricius's portrayal of the ostiolae in man. Plate II from *De Venarum Ostiolis*, in *Anatomices et Chirurgiae ... Tractatus Quatuor*, Frankfurt, 1624.

TAB XXII.

FIG. XLV.

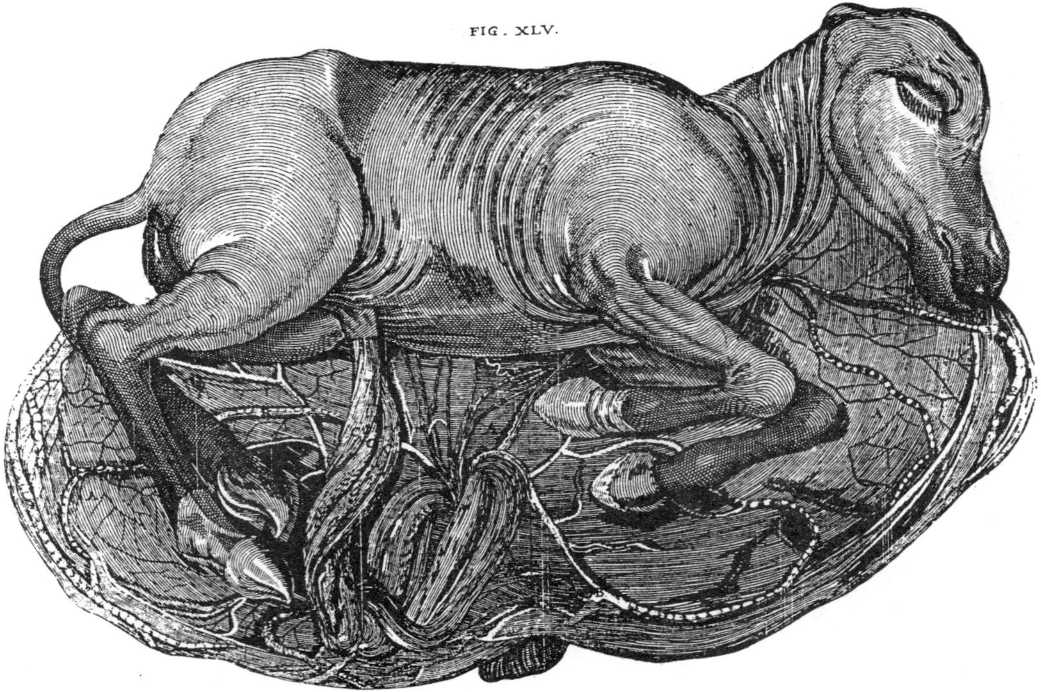

6.3 Fabricius: foetus of the horse, one of the wide range of animals used by Fabricius. Plate XXII from Fabricius's *De Formato Foetu*, in *Anatomices et Chirurgiae ... Tractatus Quatuor*, Frankfurt, 1624.

(iii)

Fabricius was the first but not the only person to pursue the anatomical project of Aristotle. He taught generations of students to see anatomy in this way, and he influenced a few of these sufficiently that they took up such anatomizing themselves.[26] Fabricius's most famous pupil was William Harvey, who returned to London from his student years at Padua and immediately took up anatomizing in the way he had been taught by Fabricius. The titles of his two published books reveal the Aristotelian nature of their themes: *An Anatomical Exercise on the Motion of the Heart and Blood in Animals* of 1628, and *Exercises on the Generation of Animals* of 1651. Both of them are about the vegetative soul: 'The heart of animals is the foundation of life, the ruler of each of them, of the microcosm it is the sun from which all *vegetatio* derives, all vigour and strength emanates,' as Harvey wrote in the Dedication to the first of these books.[27] In the Preface to the second, along with an extensive and explicit defence of Aristotle's way of gaining firm knowledge, Harvey outlined his own book on generation thus:

> We will expound firstly in the egg, then also in other conceptions of different creatures, what is constituted first and what afterwards by the great divine power of Nature with foresight, inimitable intellect and admirable order; and then we will discuss what we shall have observed about the first matter from which, the first efficient by which, and about the sequence of generation and its economy; so that thence we might grasp something for certain about each faculty of the formative and vegetative soul from its operations, and of the nature of the soul from the members or organs of the body and from their functions.[28]

Harvey also worked on other Aristotelian anatomic themes such as *respiration* and *the local motion of animals*. The project from which his discovery of the circulation of the blood resulted was one drawn up on strictly Aristotelian/Fabrician lines. For it was about 'the movement, pulse, action, use and usefulnesses of the heart and arteries' in animals – the central organ of the vegetative soul – and it was a topic available for and demanding investigation primarily because 'when Fabricius learnedly and accurately delineated in a particular treatise each part almost of animals, he left only the heart unattempted'.[29] After many years of his own experience of such anatomizing, it is no surprise that Harvey could take such research considerably further than his teacher had, and hence in his published works Harvey can be found to criticize Fabricius, and indeed Aristotle, on many occasions. But on the other hand they were, as he himself said, his only real predecessors: 'I am treading in the footsteps of other people who have borne a light in front of me in this business, and I employ their terms as much as possible. Above all others I follow Aristotle from among the Ancients, Hieronymous Fabricius ab Aquapendente from among the Moderns; I follow the former as Leader, the latter as Pathfinder.'[30]

Thus in the case of Fabricius, once more a new body is revealed to sight in the sixteenth century, and again it is an ancient body resuscitated. Each new resuscitated body resulted from a new and distinctive research programme in anatomy: that of The Physician (Galen/Vesalius) and that of The Anatomists (Herophilus-Erasistratus/Columbus), and now finally that of The Philosopher (Aristotle/Fabricius).

Notes

1. As a result of this realization historians have come to think that there were many 'Aristotelianisms' at this period, rather than a single monolithic one. See Charles B. Schmitt, *Aristotle and the Renaissance*, Harvard University Press, 1983; and his articles reprinted as *Studies in Renaissance Philosophy and Science*, London, Variorum, 1981, and *The Aristotelian Tradition and Renaissance Universities*, London, Variorum, 1984; see also Brian P. Copenhaver and Charles B. Schmitt, *Renaissance Philosophy*, Oxford University Press, 1993, esp. Chapter 2, 'Aristotelianism'. Personally, I prefer not to discuss this issue in terms of -isms.
2. This list is at Padua University; it is published in Lucia Rossetti, *The University of Padua: An Outline of its History*, Trieste, Edizioni Lint, 1983, first ed. Milan, 1972, p. 35.
3. For the teaching and the holders of the various chairs see J. Facciolati, *Fasti Gymnasii Patavii*, 2 vols, Padua, Manfre, 1757, and J.P. Tomasini, *Gymnasium Patavinum libris V comprehensis*, Udine, Schirattus, 1654. For a general account see Jerome J. Bylebyl, 'The School of Padua: Humanistic Medicine in the Sixteenth Century', in Charles Webster, ed., *Health, Medicine and Mortality in the Sixteenth Century*, Cambridge University Press, 1979, pp. 335--70.
4. John Hermann Randall, Jr., *The Career of Philosophy*, Columbia University Press, 1962, p. 74. One has to beware of interpreting all the published materials of the professors simply as evidence of fundamental differences between them, and of not noticing the profound similarities in their views.
5. 'Hac de re [viz., de norma et ratione philosophandi] verba facturus ad duo capita totam orationem meam redigam; cum enim versandum nobis sit in Aristotelis philosophia, videndum primo loco est quam philosophandi rationem servaverit Aristoteles, deinde vero quid nobis agendum sit ut in Aristotele dicamur rite et aristotelice philosophari', Mario Dal Pra, 'Una "Oratio" Programmatica di G. Zabarella', *Rivista Critica di Storia della Filosofia*, 1966, **21**, pp. 286–90, p. 287, my translation.
6. Randall, op.cit., Chapter 2.
7. Ibid., p. 84.
8. Quoted by Facciolati, op.cit., p. 274; on Cremonini, Piccolomini, Zabarella and their writings, see Charles H. Lohr, *Latin Aristotle Commentaries, II*

Renaissance Authors, Unione Accademica Nazionale Corpus Philosophorum Medii Aevi Subsidia VI, Florence, Olschki, 1988.

9. Zabarella, *Commentarii in De Anima*, 1606, as translated and cited by Randall, op.cit., p. 85. For various aspects of Zabarella's thinking (though not on the soul), see William F. Edwards, 'The Logic of Iacopo Zabarella, 1533–89', Unpublished Ph.D. Thesis, Columbia University, 1960; Antonino Poppi, *La Dottrina della Scienza in Giacomo Zabarella*, Padua, Antenore, 1972; and also Neal W. Gilbert's entry on Zabarella in *The Encyclopaedia of Philosophy*, vol. 8, p. 366.
10. Some of the issues in this section have been dealt with before in Andrew Cunningham, 'Fabricius and the "Aristotle Project" in Anatomical Teaching and Research at Padua', in A. Wear, R.K. French and I.M. Lonie (eds), *The Medical Renaissance of the Sixteenth Century*, Cambridge University Press, 1985, pp. 195–222, from which I borrow freely. G. Favaro, 'L'Insegnamento Anatomico di Giralomo Fabrici d'Aquapendente', *Monografie Storiche sullo Studio di Padova*, Venice, 1922, deals only with the institutional aspects of Fabricius's teaching, not with its content.
11. Officially the old statutes still governed the anatomy demonstrations, with one professor of medicine obliged to read the text of Mundinus while another commented and showed the parts, but Fabricius does not seem to have considered himself particularly constrained by these statutes. The statutes operative during Fabricius's professorship are in *Statuta Almae Universitatis D. Artistarum, et Medicorum Patavini Gymnasii. Denuo Correcta, et Emendata*, Venice, Painus, 1589, and *Statuta Almae Universitatis DD. Philosophorum, et Medicorum Cognomento Artistarum Patavini Gymnasii. Denuo Correcta, et Emendata, et Nonnullis Apostillis Scitu Dignis Aucta*, Padua, Rosatus, 1607; see Book 2, c. 28, which is identical in both. Facciolati, op.cit., vol. 2, p. 388, points out that Fabricius had the permission of the Venetian Senate to both dissect and show the parts himself.
12. There had been an earlier permanent theatre built in 1584, which was put up in a hurry to tempt striking students back to Padua, and this was replaced in 1594 by the theatre which still survives. On which see the forthcoming work of Richard Palmer in *History of Universities*; and Tiziana Quirico (ed.), *Les Siècles d'Or de la Médecine: Paduoue XVe-XVIIIe siècles*, Milan, 1989, (Catalogue of an exhibition at the Jardin des Plantes, Paris, 1989, organised by Pietro Corsi), pp. 106–9; M. Alain Segonds of the C.N.R.S., Paris, kindly acquired this catalogue on my behalf; E. Ashworth Underwood, 'The Early Teaching of Anatomy at Padua, With Special Reference to a Model of the Padua Anatomical Theatre', *Annals of Science*, 1964, **19**, pp. 1–26.
13. See, for instance, A. Favaro, *Atti della Nazione Germanica Artista nello Studio di Padova*, 2 vols, Venice, 1911, vol. 1, p. 286 for an instance in 1590; and for a general consideration of these conflicts, see Howard B. Adelmann, *The Embryological Treatises of Hieronymus Fabricius of Aquapendente*, 2 vols, Cornell University Press, 1942, republished 1967, vol. 1, pp. 12–22.
14. See above, note 10.
15. See the Dedication to *De Aure*, 1600.
16. There are, however, certain lectures of Fabricius on 'Anatomical Method' preserved at the Biblioteca Communale, Siena.

17. *De Methodis*, cited by Randall, op.cit., p. 298.
18. Dedication to *De Formato Foetu*, Venice, 1600, reprinted in facsimile in Adelmann, op.cit., vol. 2, p. 464, my translation, based on that of Adelmann, ibid., vol. 1, p. 238.
19. For getting a sense of what kind of work and mental switch might be involved in putting an Aristotelian project into renewed practice it is valuable to see the writings of some present-day Aristotelians: John Wild, *Introduction to Realistic Philosophy*, New York, Harper Brothers, 1948, and Francis H. Parker and Henry B. Veatch, *Logic as a Human Instrument*, New York, Harper Brothers, 1959.
20. The following account is built up from what Fabricius says in different places in his writings, see especially *De Visu*, 1600.
21. On which see N. Jardine, 'Galileo's Road to Truth and the Demonstrative Regress', *Studies in History and Philosophy of Science*, 1976, 7, pp. 277–318.
22. My translation; for an alternative translation see K.J. Franklin, *De Venarum Ostiolis 1603 of Hieronymus Fabricius of Aquapendente*, Springfield, IL, Thomas, 1933.
23. In his *The Life of Father Paul, of the Order of the Servie. Translated out of the Italian by a Person of Quality*, London, Moseley and Marriot, 1651, pp. 30–31, Fulgenzio Micanzio claimed that his hero, Friar Sarpi, had predicted the finding of the *ostiola*: 'considering the gravity and weight of the blood [Sarpi] grew into opinion that it could not stay in the veins, except there were some bunch to hold it in, some folds or shuttings, at the opening and closing of which there was given a passage and necessary Equilibrium to life. And upon his own natural judgement he set himself to cutting with more exquisite observation, whereupon he found out those *Valvulae*, and the right use of them, which do not only stop and hinder the blood from dilating itself by his weight into the veins . . . but also that blood running up and down with so much liberty, and in so great quantity it might easily suffocate the natural heat of those parts which ought to receive their nourishment from it . . . Of this he gave account to some friends of that profession, and especially to Aquapendente, who made very great use of it in a public Anatomy, after which divers authors have written much upon that subject'. It is not necessary or desirable for us to adjudicate between the claims of Fabricius and Sarpi.
24. Dedication to *De Visu*, 1600.
25. See G. Sterzi, 'Le "Tabulae Anatomicae" ed i Codici Marciani con Note Autografi di Hieronymus Fabricius ab Aquapendente', *Anatomischer Anzeiger*, 1909, 35, pp. 338–48. The printed catalogue of St Mark's Library is not very informative; see Joseph Valentinelli, *Bibliotheca Manuscripta ad S. Marci Venetiarum*, Venice, 1868, p. 64 (where the date of Fabricius's death is given as 1609). Some of the paintings have been published; there are two in colour in Gweneth Whitteridge, *William Harvey: Disputations Touching the Generation of Animals*, Oxford, Blackwell, 1981, and eight in colour in *Les Siècles d'Or de la Médecine*.
26. Students visiting several universities would, of course, be exposed to several ways of anatomizing and might well make an unconscious amalgam of them all if they came to practise anatomy themselves. But Julius Casserius (lecturer at Padua 1609–16) seems to have conducted such Aristotelian anatomy, to judge by his printed works. In his Dedication to

his *De Vocis Auditusque Organis Historia Anatomica*, Ferrara, Baldinus, 1600–01, he writes, 'Si quae vero admiratione digna Naturae habet contemplatio, ea profecto sunt quae ad animae facultates earumque instrumenta pertinent' (f. a.2). Similarly his *Pentaestheseion, Hoc est de Quinque Sensibus Liber*, Venice 1609 (reissued as *Nova Anatomia*, Frankfurt, 1622) was an Aristotelian piece of anatomy. In true Aristotelian mode, Casserius is concerned with 'the sensed' as well as with the sense organ itself. His successor as Paduan anatomy teacher, Adrianus Spigelius, however, is a Vesalius-style anatomist in his *De Humani Corporis Fabrica Libri Decem*, Frankfurt, 1632 (posth.), and even in his *De Formato Foetu Liber*, Padua, 1626 (posth.) which deals entirely with the human fetus. Spigelius was from Brussels (like Vesalius) and though Padua-educated had spent many years in the north of Europe before becoming professor at Padua.

27. 'Cor animalium, fundamentum est vitae, princeps omnium, Microcosmi Sol, a quo omnis vegetatio dependet, vigor omnis & robur emanat', *Exercitatio Anatomica de Motu Cordis et Sanguinis in Animalibus*, Frankfurt, Fitzer, 1628, p. 3, my translation.
28. 'Nos igitur (secundum methodum a nobis propositam) primum in ovo, deinde in aliis quoque diversorum animalium conceptibus, quid prius, et quid posterius a summo naturae numine, cum providentia, et intellectu inimitabili, ordineque admirabili constituatur, exponemus; simulque tandem, de materia prima ex qua, et efficiente primo a quo, deque generationis ordine, et oeconomia, quae observata a nobis fuerint, enarrabimus; ut inde, de quacunque formatricis, et vegetantis animae facultate, ex operibus ipsius; de animae natura, ex membris sive organis, eorumque functionibus, aliquid certi intelligamus', *Exercitationes de Generatione Animalium* (originally published London, 1651), as reprinted in *Opera Omnia*, London, Royal College of Physicians, 1766, p. 178, my translation.
29. 'Sed et eo lubentius, quod Hieronym. Fabr. ab aq p. cum singulas pene animalium particulas, accurate, et docte peculiari tractatu delineaverat, solum cor intactum reliquit', *Exercitatio Anatomica De Motu Cordis et Sanguinis in Animalibus*, Frankfurt, Fitzerus, 1628, p. 21, my translation. See also Andrew Cunningham, 'William Harvey', in Roy Porter, ed., *Man Masters Nature*, London, BBC, 1987, pp. 65–76.
30. 'Id ne fieret [ie coin new terms], aliorum, qui in hoc negotio mihi facem praetulerunt, insisto vestigiis; eorumque (quoad licuit) utor verbis. Prae caeteris autem, *Aristotelem* ex antiquis; ex recentioribus vero Hieronymum *Fabricum ab Aquapendente*, sequor; illum, tanquam *Ducem*; hunc, ut *Praemonstratorem*'. Preface of Harvey's *Exercitationes de Generatione Animalium*, London, Pulleyn, 1651, my translation.

Part II
The Anatomical Reformation? An Enquiry

What a piece of work is a man! How noble in reason! how infinite in faculty! in form, in moving, how express and admirable! in action how like an angel! in apprehension how like a god! the beauty of the world! the paragon of animals!

Shakespeare, *Hamlet*, Act 2, Scene 2

CHAPTER SEVEN

An Anatomical Reformation?

(i)

In pursuit of answers to my basic questions – what were these anatomists doing? and why were they doing it? – this is how I have so far characterized the 'what'. First, these sixteenth-century anatomists were seeking to emulate the anatomical projects of certain Ancients. Second, this led them to anatomize in different ways than hitherto, and hence changed the mode and nature of their anatomizing. Third, it gave them different 'spectacles' with which to see the body and hence led them to see hitherto unsuspected things to be true about the body. Thus Vesalius emulated Galen, Colombus emulated Herophilus and Erasistratus, Fabricius emulated Aristotle. As a consequence of such adoptions of the anatomical projects of particular Ancients, Columbus, for instance, was able to discover the so-called lesser circulation, and William Harvey, Fabricius's pupil, was able to discover the circulation of the blood.

But why were they doing this? Obvious as it may be to turn to this question, before we embark on this difficult terrain we need first to note that explicit discussion of causal questions has rarely much concerned historians of science or medicine, and they generally assume that the accounts they give are indeed causal, even though there is no worked-out agreement as to what does, and what does not, count as a properly causal account.

Returning to the particular question about why this variety of anatomical projects was revived in the sixteenth century and put back into practice, I have so far (and this has been deliberate policy on my part) used only one category to account for anatomists taking up and pursuing their particular anatomical projects: the Renaissance. 'Emulation of the Ancients' is an essential, identity-giving feature of the Renaissance as that concept is commonly understood, and by seeking to emulate the Ancients in their anatomizing, these Moderns were evidently typical Renaissance figures, and this accounts for why they were doing what they were doing. Thus although what I have said about the content and form of the anatomizing of these sixteenth-century anatomists is new, yet the structure of my account of them and their innovations nevertheless fits well within conventional understandings of the Renaissance.

For some people this may be a pefectly satisfactory explanation. And up to now I have deliberately treated the Renaissance as an unproblematic category, even using it for the title of the present book. But of course it is not. And this fact presents particular difficulties for the history of anatomy. I am not referring here to the difficulty of agreeing temporal boundaries to the Renaissance, though these trouble historians still, nor to the difficulty of distinguishing between Renaissance phenomena and the medieval phenomena they supposedly supplanted. The problems lie in the very definition and understanding of the Renaissance that we use. For while the term Renaissance does conform to some fifteenth and sixteenth-century views of what was happening, with certain people feeling they were living through an age of renewal, and the term itself actually being coined in the period ('*Rinascita*', coined by Vasari in 1550), *our* concept of the Renaissance derives not from the fifteenth and sixteenth centuries but from the nineteenth century. As Wallace Ferguson has conclusively shown,[1] the modern-day concept of the Renaissance, both as a period and as a phenomenon, derives primarily from Jacob Burckhardt's book of 1860, *The Civilization of the Renaissance in Italy* (*Die Cultur der Renaissance in Italien*). Burckhardt in his turn was building on the vision of Stendhal, Jules Michelet and other early nineteenth-century writers.[2]

For the essence of Burckhardt's (and thereby our) concept of the Renaissance we may take a definition put forward by his follower, John Addington Symonds. The Renaissance, Symonds claimed, was

> a comprehensive movement of the European intellect and will toward self-emancipation, toward reassertion of the natural rights of the reason and the senses, toward the conquest of this planet as a place of human occupation, and toward the formation of regulative theories both for states and individuals differing from those of medieval times.[3]

Or again,

> what the word Renaissance really means is new birth to liberty – the spirit of mankind recovering consciousness and the power of self-determination, recognizing the beauty of the outer world, and of the body through art, liberating the reason in science and the conscience in religion, restoring culture to the intelligence, and establishing the principle of political freedom.[4]

Symonds's bold way of writing is somewhat over the top by modern standards, and few if any historians would today accept these definitions in their totality, yet all attitudes toward or against the Renaissance are still today shaped by such assumptions; indeed, one cannot talk about the Renaissance at all without talking in (or implicitly accepting) such terms, for they are the terms which give the concept its current meaning. Thus even the attempts on the part of historians who study the medieval period to question or challenge the importance of the

Renaissance – by showing that it had been preceded by earlier 'renaissances' (such as one in the twelfth century), and that all its vaunted achievements had already been achieved before the sixteenth century – still use *the same definitional categories* of what constitutes a proper renaissance. Hence such rival versions of what constitutes an *earlier* renaissance, or arguments as to why the sixteenth-century Renaissance was not a proper renaissance after all, are themselves necessarily framed in Burckhardtian terms.[5]

What are these definitional categories? As Ferguson has shown, the fundamental Burckhardtian concept of the Renaissance, the concept which we have inherited, was constructed out of typically nineteenth-century elements and reflected typical nineteenth-century concerns.[6] In the first place there was the Romantic concept of the development of the individual. Burckhardt speaks of it thus:

> In the Middle Ages both sides of human consciousness – that which was turned within as that which was turned without – lay dreaming or half awake beneath a common veil. The veil was woven of faith, illusion, and childish prepossession . . . Man was conscious of himself only as a member of a race, people, party, family, or corporation – only through some general category. In Italy this veil first melted into air; an *objective* treatment and consideration of the State and of all things in the world became possible. The *subjective* side at the same time asserted itself with corresponding emphasis; man became a spiritual *individual* and recognised himself as such. In the same way the Greek had once distinguished himself from the barbarian . . .[7]

The apogee of the spirited individual was the 'genius', the 'all-sided man' or 'l'uomo universale': the 'artists who in every branch created new and perfect works, and who also made the greatest impression as men', and some of them 'who may truly be called all-sided, tower above the rest'.[8] This distinction between the subjective and the objective, this belief in the importance of individuality and its cultivation, this belief that certain rare individuals have creative individuality in super-abundance and deserve celebration on that account, these are all nineteenth-century Romantic attitudes being projected back onto the fifteenth and sixteenth centuries. The case is the same in Burckhardt's and his followers' view of the Renaissance as simultaneously the product and the expression of the Spirit of the Age engaged on its 'self-realization'. Here it is Idealistic (Hegelian) attitudes which are being projected back from the nineteenth to the fifteenth and sixteenth centuries.[9]

Secularization and liberalism are further nineteenth-century attitudes which were being projected back on the fifteenth and sixteenth centuries by Burckhardt, and which thus became definitional categories of it:

> These modern men, the representatives of the culture of Italy, were born with the same religious instincts as other medieval Europeans. But their powerful individuality made them in religion, as in other matters, altogether subjective, and the intense charm which the discovery of the inner and outer universe exercised upon them rendered them markedly worldly . . . their belief in God began to waver, and their view of the government of the world became fatalistic . . .[10]

The Renaissance was thus conceived and presented by Burckhardt and his followers as a freeing of mankind from the superstition of medieval times, as a free-thinking (possibly even pagan) event liberating men from the constrictions of authoritarian Christianity. The corollary of this, according to Burckhardt, was the rejection of religious authority in favour of trusting *the self*, the recognition of the capacity and creative potential of the human spirit and the attempt to seek its fulfilment or self-realization in free action and enquiry. It was thus the period of 'induction and free inquiry', the very attitudes in which the Middle Ages had been deficient.[11] Hence their decision to bind themselves to the emulation or imitation of Antiquity was not actually a new bondage for Renaissance figures but a liberation, for in doing this they learnt to behave just like the Ancients themselves and thereby come to trust in *Nature* herself. As Burckhardt put it:

> culture, as soon as it freed itself from the fantastic bonds of the Middle Ages, could not at once and without help find its way to the understanding of the physical and intellectual world. It needed a guide, and found one in the ancient civilization, with its wealth of truth and knowledge in every spiritual interest. Both the form and substance of this civilization were adopted with admiring gratitude; it became the chief part of the culture of the age. The general condition of the country was favourable to this transformation.[12]

'The spirit of the people, now awakened to self-consciousness, sought for some new and stable ideal on which to rest'.[13] It found it in emulation of the Ancients. And all good nineteenth-century Romantics and Idealists knew that the Greeks had been the very models of individuality, who had ignored 'rules' – which simply stifle creativity – and had simply imitated Nature alone. As Hegel put it, the Greek spirit had been 'Individuality conditioned by Beauty', and this is (supposedly) what the men of the Renaissance were imitating.[14] And as the art historian Erwin Panofsky has commented more recently, 'To Petrarch's theme, "*back to the classics*", Boccaccio and Villani (followed by many other writers on painting . . .) opposed, as a counterpoint, the theme, "*back to nature*"'.[15]

Historians of other Renaissance topics may or may not find any problems for their practice in the continued dominance of this nineteenth-century historiography. But there are three main problems for our history of anatomy raised by the use of the concept Renaissance, defined in this nineteenth-century way – and it must, as we have seen, unavoidably be defined in this way.

First, 'cause' is excluded from discussion *by definition*, except insofar as 'the Spirit of the Age' may be said to have been 'realizing' itself or exceptional men were 'realizing' themselves; the Renaissance, Michelet claimed, 'started from nothingness. It was the heroic outburst of an immense will'.[16] The whole tradition of *Kulturgeschichte*, of which the concept of the Renaissance has been a part since the nineteenth century,[17] separates the intellectual and the artistic achievements of man from their economic and social matrix, treats them as being, by their very nature, inexplicable by economic or social causes, but only in terms of the strivings of the human spirit. For the historian of culture, art is autonomous and far too important to be caused by economic conditions; the world does not mould the individual, the individual moulds the world (and this is another typical Romantic attitude thrown back onto the fifteenth and sixteenth centuries). In other words, the concept of the Renaissance leads the historian to treat the effects (the achievements of the mind) as being either without cause or as being the causes of themselves. Certainly they cannot be accounted for or explained in terms of their causes. Yet we only have to think about some of the people conventionally treated as stars of the Italian Renaissance, people such as Donatello, Brunelleschi, Leonardo da Vinci, Michelangelo or whoever, to realize that all of them either were born or educated or pursued their adult lives in one small area: the northern part of Italy. Every 'genius' of the Italian Renaissance could be plotted onto a map of north Italy. What brought so many talented and very able men to the fore was not some sudden outbreak of monstrous births, nor some magic in the water. Rather, it was of course the economic and social circumstances of that region at that precise time. North Italy was quite simply the richest and most commercially active area of the whole of western Europe. In the decades around 1500 the city-states of north Italy peaked in population, in commercial enterprise, in urban growth. It was at this time that the economy of the north Italian states first regained and then passed the heights it had reached at the time of the Black Death of 1348. Economic recovery was complete, and economic growth was still soaring.[18] The artistic and intellectual flourishing that we detach from these economic and social events by our use of the concept of the Renaissance, was a particular side-effect of the surplus wealth and surplus energy of the people of this region in this condition. The economic and social circumstances were now such that talented local youngsters could be given an opportunity to realize their abilities in raising buildings, painting pictures, sculpting figures, and young hopefuls from beyond the Alps could move into the area too in the hope of being able to sell their talents to the rich Italians. I do not want to offer an opinion here on the nature of the detailed causal connections which linked these economic and social conditions to the production of Renaissance works of the intellect and the imagination; I simply want to point out that our present concept of the Renaissance denies that there are any links at all, beyond the relatively trivial one of patronage. Thus for the historian of anatomizing the concept of the Renaissance is of negative use in trying to explain *why* it was in this particular place and time that new anatomical projects were first engaged in.

The second problem that the use of this nineteenth-century concept of the Renaissance raises in the history of anatomy, is that it gives no account at all of why the Moderns sought to emulate the Ancients in the first place. Yet this is the most fundamental feature of the concept of the Renaissance: that it was a *re*-birth. Why did the Moderns look to the past rather than to the future: to the dead rather than the living? No historian to my knowledge has tackled this most basic of questions about the Renaissance, and I have no special answer to it either. But until it is tackled, the Renaissance must remain a basically circular concept: people who sought to revive ancient practices were, by definition, Renaissance figures; and the Renaissance was, again by definition, the phenomenon of people trying to revive ancient practices! The question 'why?' simply has no purchase on this way of conceptualizing the Renaissance. The historian of anatomy is therefore unable, as long as he uses this concept, to investigate *why these anatomists sought their inspiration in the distant past.*

The third and most important problem for the history of anatomy, is the way in which in the Burckhardtian tradition the Renaissance is taken to have been an essentially *secularizing* movement. The problem lies in the fact that, as has already become clear in my earlier accounts of both ancient and revived anatomy, anatomy in the western tradition was essentially about the soul. And as there has been no other cultural tradition apart from the western one which has systematically dissected the dead to understand the functioning of the living, this means that *all* systematic anatomizing was and had been about the soul. Every anatomist up to and well beyond the sixteenth century looked at the body as being, in one way or another, the instrument of the soul, and if they were interested in anatomizing it, it was *because* it was the instrument of the soul. And the soul is perhaps the most unsecular thing one can think of, since the central concern of the whole of organized religion in the West, Christianity, is and always has been with the nature, the career and the fate of the soul. The soul is what religion is about in Christendom. And the soul is what anatomizing was about.[19]

So if we are to make sense of the history of anatomy in the sixteenth century, with its central concern with the soul, we cannot appeal to a concept of the Renaissance which sees it as a secularizing movement, as a movement which stepped aside and stood aloof from Christianity and matters of religious faith – and which succeeded because it did so. The present understanding of the concept of the Renaissance still has secularization and liberalism at its core, and as this concept of it is so widely held, with whole university departments founded upon it and with the past periodized in terms of it, it would be a bold person indeed who tried singlehandedly to argue for a concept of the Renaissance which was religious through and through.[20] We need to turn elsewhere.

(ii)

It might be thought that we could find appropriate explanatory categories in the discipline of *the history of science*, especially since the history of anatomy is generally seen as a part of the history of science and medicine. But in turning to the history of science, we need to constantly bear in mind that any account of sixteenth-century anatomizing which omits or downplays the role of the soul is missing the point. As virtually every account of the history of anatomy that we currently have is of this kind, it does, of course, make sixteenth-century anatomy sound more like twentieth-century anatomy, from which consideration of the soul is totally absent. But that is just what I mean: it misses the point. And this is precisely what the history of science, as conventionally practised, does. That is why we cannot appeal for suitable explanatory categories to the history of science, for concern with the soul is as alien to the history of science, as conventionally conceived and practised, as it is to the concept of the Renaissance: the definitional categories of the history of science preclude the soul from being the central concern of sixteenth-century anatomy. It is worthwhile seeing why this is the case.

Historians of science and medicine have been generally very unconcerned about the origins of their own discipline and the values it embodies and promotes and have tended to assume that the practice of science is of great antiquity, and that writing its history (the finding of episodes of the practice of science in the past, and recounting them) is pretty unproblematical, especially for someone who has modern scientific training. But the shape and concerns of the conventional account of the history of science and medicine were established in the nineteenth century, and reflected the concerns of nineteenth-century men, and this fact is very important for our present concerns. For the present-day concept of the history of science and medicine is thus the counterpart of the concept of the Renaissance that we have already dealt with. For it was constructed, just as in the case of the concept of the Renaissance, by projecting back nineteenth-century concerns onto earlier centuries, including the sixteenth and seventeenth. And, given that this history of science was the counterpart of the history of art, to which the concept of the Renaissance was so central, it is not surprising that the concerns which went to constitute the identity of the history of science were mostly the same nineteenth century Romantic, liberal and Idealist concerns which went to constitute the concept of the Renaissance.[21] There was the Romantic and liberal concept of 'individuality': the view of the individual as one engaged on a personal journey of intellectual 'self-realization', who seeks knowledge and insight simply 'for its own sake'. As Burckhardt himself put it:

> we hold that, at any time, and among any civilized people, a man may appear who, starting with very scanty preparation, is driven by an irresistible impulse into the path of scientific investigations, and through his native gifts achieves the most astonishing success.[22]

There was the Romantic and liberal valuation of 'freedom of thought' as a supreme good, and equally there was the contempt for people who believed what authority told them rather than thinking for themselves. There was the Romantic concept of the 'genius', the supreme creative thinker who breaks all the rules and fights the intellectual battles against the forces of reaction and stupidity, and who is therefore a hero. There was the Romantic concept of the distinction between the subjective and the objective, with the subjective now in the nineteenth century appropriated to be the realm of art and the objective to be the realm of science. The people who constructed this account of the past of science were primarily propagandists for science, and those of them who were also Idealists or Hegelians also saw science and its individual doctrines as having an in-built *internal development* which unfolds over the centuries, in accordance with the actions of the 'Spirit of the Age'. And both the Idealists and the positivists thought that *progress* was of the essence of science, and that the historian of science should therefore seek out and show the progress of knowledge over the centuries.[23]

Just as in the case of the concept of the Renaissance, so one of the pre-suppositions of that history of science which was created in the nineteenth century, and which has been developed in the twentieth century, was that the *secularization of knowledge over time* was one of its most fundamental features: when religion dominated men's minds, they saw Nature only through a glass darkly, but when they recognized the proper distinction between religion and the inquiry into Nature, they saw Nature face to face and then science began to flourish. Thus what was presented by the nineteenth-century people who were first writing the history of science (and who were primarily doing so as propagandists for science) was the story of an age-old conflict between science and religion; this story has continued to be told by historians in the twentieth century. In this story, the free-thinking Greeks pursued science successfully; but the men of the Middle Ages could not do so since their eyes were blinkered by the demands of their religion. Only when they realized that the domain of religion and the domain of the exploration of Nature are quite distinct, could they see Nature clearly again, and then science flourished once more. This latter turning-point, this turning back to Nature, happened (according to this story) during the Renaissance, for that was the period when the Moderns learnt again to read Nature without preconceptions and without deference to authority, simply trusting in their own capacity for observation, and they did so by imitating the approach of the ancient Greeks. We can see here how, under this interpretation, for our Renaissance anatomists the emulation of Antiquity – the re-adoption of various ancient anatomical practices – can be seen as them trusting their own eyes, discarding the dictates of 'authority', whether that of human authors or of the Church, and simply turning to Nature herself. If imitating the Greeks led them to imitate Nature herself, then their Renaissance anatomical investigations were successful and productive of new knowledge because they were achieved by throwing off the traditional deference to Christian teaching.

This assumption of the fundamental incompatability between religion and the successful pursuit of knowledge about Nature still provides one of the basic structures for the discipline of the history of science in the twentieth century. Indeed, according to many twentieth-century historians of science, this new attitude of open-minded inquiry toward Nature from the sixteenth century onwards had such enormous consequences that it constituted nothing less than a – indeed *the* – 'Scientific Revolution': the revolution from the scholasticism and Christian deference of the medieval period back to the science of the Greeks, and then the making of advances on that revived Greek science. Professor Rupert Hall, who may be taken as speaking for a generation of historians of science, puts it like this in introducing his own book called *The Revolution in Science 1500–1750*. Before this period, Hall writes,

> above and beyond all forms of authority was religious truth ... Even the attempt to distinguish a natural event from a divine intervention involved theological considerations: and there could be nothing of *science* until such a distinction had been drawn ...

> this book might be said to be about the victory of rationality over religiosity.[24]

And in introducing a work called *The Scientific Renaissance*, he wrote:

> *The Scientific Renaissance* describes the early stages of the Scientific Revolution, beginning with what is traditionally (but somewhat inaccurately) known as the Renaissance of Learning in the fifteenth century. The Scientific Revolution was the effect of a unique series of innovations in scientific ideas and methods; it gave the key to the understanding of the structure and relations of things. It was (and still remains) the greatest intellectual achievement of man since the first stirrings of abstract thought, in that it opened the whole physical universe – and ultimately human nature and behaviour – to cumulative exploration . . . [It was part of the European intellectual tradition] And this, of course, was founded in Greece. The Greek philosophers, imposing no bounds on intelligence but those of the universe itself, set at the very root of the European tradition of science the ideal of an interlocking system of ideas sufficient to explain all the variety of nature . . . Although the Scientific Revolution ultimately came as a reaction against the dogmatism inherent in the emulation of antiquity, it too drew its inspiration from neglected aspects of the Greek legacy . . . The Scientific Revolution did not reject Greek science; it transformed it . . .[25]

That has been the general view of the history of science as it has been cultivated since the nineteenth century, built on and still promoting the nineteenth-century values of liberalism, anti-clericalism, objectivity, individuality and so forth. But we need to remember that this skeleton outline of the history of science as an enterprise which has been engaged in spasmodically since Greek times, flourishing when religiosity is held at bay, was the construction initially of people who were not historians of science (there was no such class of men) but propagandists for science, and it was cobbled together for propaganda purposes rather than being the outcome of an extensive and dispassionate inquiry into the historical sources. It was this propagandist story which was taken up by the handful of professionals – the academics – who then pursued the history of science over the next half century or so; and they introduced a number of modifications into that stark story, especially by reassessing in more positive terms the 'contribution' that the men of the (supposedly benighted) medieval period had made toward the development of science. Then, over the last few decades, a fair-sized profession of historians of science has come into existence. It is no surprise that this little army of historians has found many inconsistencies and ambiguities in this jerry-built nineteenth-century story of the history of science – especially as they themselves have often been imbued with some different values, the values of the 1960s, 1970s and 1980s, when the claims of science were being widely questioned. Among other avenues they have opened up have been questions about causation, motivation and intentionality in certain of the celebrated moments of the history of science, and they have applied sociological and even anthropological approaches to assessing many of those moments. In particular a number of them have reassessed the relationship of religion to science in the past, and have been willing to find that the science of, for instance, Paracelsus in the sixteenth century or Robert Boyle or Isaac Newton in the seventeenth, was in fact *heavily religious* in aim and content. Thus it is currently accepted by historians of science of the present generation that there was really no essential incompatability between the science and religion of people in the sixteenth and seventeenth centuries, and that in practice the science was highly religious in goals and content, or that their science was actually religious (or magical or mystical or in some other way non-scientific).[26]

Such new explorations have produced many changes, and we now have a much more nuanced reading of the history of science, with the sharpness of the old categories much softened. But modern scholars have not got themselves a new reading of the *whole* of the history of science – no new big picture – with which to replace that nineteenth-century construct.[27] So, just as in the case of those twentieth-century historians who have opposed certain aspects of the nineteenth-century concept of the Renaissance, modern historians of science make their reassessments still *from within* a nineteenth-century framework of thinking. Hence, despite all the significant changes which modern historians have indeed brought about in the story of science's history, their account of it still has pretty much the same temporal range, pretty much the same periods of great flourishing, pretty much the same roll-call of heroes with much the same achievements held to their respective credits, pretty much the same strongly approving attitude to instances of apparent free and independent thinking among past explorers of Nature. They still people the past with 'scientists' (although this concept was not current before 1833, when the term was coined) engaged on early versions of present-day questions about Nature, and they still trace cumulative, progressive histories for the various scientific disciplines. And when they discuss the relation between science and religion in the past, historians of science today offer an account which is still basically about their opposition, but now they argue about *degree* (about *how* opposed science and religion were), some claiming that there was more (or less) religion in the science than we used to think, others that the opposition between science and religion was not so great as hitherto represented, or that there was often collaboration and compatability between science and religion rather than hostility, or even (and this is a relatively radical view still) that the science of past men of science was actually given its shape by their religious attitudes. But what none of the modern historians of science has done, as far as I can see, is put aside the science/religion polarization of the discussion about past investigation of Nature. Yet we have this polarization only because we have inherited it; the thought that the two must be enemies was an obsession of certain men of the nineteenth century. It does not obsess men of the late twentieth century so much (as the current revisionist account of the old history of science story shows, by demonstrating that they may not in fact have been incompatible). And this polarization of religion versus the investigation of Nature may well not have obsessed men of the *sixteenth* century either; indeed, it may never have occured to them. In other words, we need to ask ourselves whether the distinction is pertinent, or even relevant, to the past people whose investigations of Nature we are exploring. If we take the distinction with us to our historical inquiries into the Nature-investigations of past men, then we will of course be able to separate out into two piles, as it were, those of their views which, *to our eyes*, count as religion and those which, *to our eyes*, count as science respectively, and hence we will come to think that the evidence proves the validity of our initial categories. But if we choose not to take this distinction with us to our inquiries (and the choice of whether we do or not is entirely ours, it does not depend on the historical material itself), then we might find that the men of the sixteenth century did not take this distinction to their inquiries either.

This is the step we must take here. The first part of this book has revealed how all sixteenth-century anatomizing was built around the soul; the body was the soul in action, and anatomizing was a means to discover and admire the faculties and operations of the soul, in ever greater detail. This soul-obsession of sixteenth-century anatomizing obliges us therefore to look at such anatomizing in a new light. We will not be dealing with science on the one hand, and religion on the other, nor even with mixtures of the two in various proportions; but with an enterprise of investigating Nature which, unlike science today, was itself *religiously informed and constituted.* Therefore if we are to make sense of sixteenth-century anatomizing in its own terms, we have to put aside this central plank of the conventional view of what constitutes the shape, the concerns, and even the point of the history of science and medicine. Obviously this will be quite a big step.

(iii)

So, in search of an appropriate understanding of the history of anatomy in this period, I want at this point to turn in a different direction, and to a category of historical explanation of the sixteenth century to which the soul and religion are central: the Reformation.

What I shall present can not remotely be a taken as a comprehensive account of the significance of the Reformation for the history of anatomizing, nor as an account intended to totally displace the concept of the Renaissance in that history. It is meant instead simply to open up a possible field of inquiry, and see what light this might throw on a limited 'why?' question: *why* particular *anatomists sought to revive* particular *ancient projects of anatomizing in this period*. The ground on which I turn to religion here is that it takes us into a domain of motivation. For the adoption of a particular creed by a group or an individual, or the presence of a particular creed within a community (even if all do not subscribe to it) brings with it an enormous ideological charge. It tells people how to behave, it pervades their outlook on the world in every respect. It even tells one what is the value (if any) of looking at the contents of the world – that is, at Nature. Reciprocally, of course, it is usually because one belongs to a particular class in a particular place and time that one adopts or embraces a particular religious outlook, whether one is continuing to practice the religion of one's forefathers or whether one rejects it in favour of a new interpretation of doctrine and discipline which better matches one's economic and social position and aspirations.

In practice what I shall be looking for will be parallels (or homologies) between certain important moments of new anatomizing – those dealt with in the first part of this book, plus two others – and certain significant moments of the Reformation. The implication of my approach will be that the turning-points in the history of anatomizing in this period – the moment of turning to a new Ancient, the choice of which Ancient to turn to – was related to, indeed sparked off by, changes in the contemporary religious world and, in some cases, by changes in the confessional attitudes and allegiance of the anatomist himself. By proceeding in this way I seek to take seriously the coincidence in time, place, and in one instance even of person, of the Europe-wide changes in religious confession on the one hand and the changes in modes of anatomizing on the other. But I stress that this is a limited inquiry, one aimed simply at indicating *the sort of association* that there may have been between anatomizing and religious commitment in the Reformation. At all events it might bring us a little closer to an answer to a question a recent historian of anatomy raised when saying that we should be asking larger questions than hitherto in the history of sixteenth century anatomy: 'Why was anatomy important for the Renaissance', he asked, 'and what can account for its popularity?'[28]

Before we can get started on this inquiry into whether there was an 'anatomical Reformation', and whether this is a more appropriate description than the 'anatomical Renaissance', it is necessary to discuss the relation that historians believe obtained between the Renaissance and the Reformation.[29] And here we can say that there have been a variety of views. Some historians see and have seen the two movements as independent of each other, the one being about art and suchlike, the other about religion. John Addington Symonds, for instance, wrote of the 'Reformation' and 'Counter-Reformation' that

> these terms indicate moments in the whole process of modern history which were opposed, each to the other, and both to the Renaissance; and it is needful to bear in mind that they have, scientifically speaking, a quite separate existence [from the Renaissance].[30]

Other historians have seen them as related, such as Jules Michelet in the nineteenth century who saw the Reformation as fully part of the Renaissance, saying that the Renaissance was about the discovery of the world and the discovery of man –

> from Columbus to Copernicus, from Copernicus to Galileo, from the discovery of the earth to that of the heavens. Man refound himself. While Vesalius and Servetus revealed life to him, his moral mysteries were penetrated by Luther, by Calvin, by Dumoulin and Cujas, by Rabelais, Montaigne, Shakespeare, and Cervantes. He plumbed the profound depths of his nature.[31]

Historians after Michelet and Symonds have tended to pursue one line or the other. However, even those historians who believe these were two closely related movements, tend in practice to deal with them separately and, if they are professional historians, to work in different historical departments of universities.

But everyone will admit that both the Reformation and the literary side of the Renaissance were built on humanism and on the work of humanists, and in both cases the most important figure they would name would most probably be Erasmus, and hence the most important humanist figure was a leader in both movements. Other sixteenth-century men who were involved in fighting for or against religious reform have and had independent historical reputations as humanists: such are Melanchthon, the confidant of Luther, and Aleander who taught Greek in Paris and then opposed Luther on behalf of the pope.[32] Thus, the fact that someone may have been a humanist does not in any way indicate what his religious preference or predisposition might have been: there are humanists on both sides of all Reformation disputes. But with respect to the ambition of *going back to the Ancients*, it must be admitted that it was fundamental to those humanists engaged in both the Renaissance and the Reformation. All the famous men of the Reformation were trying to recover authentic ancient texts (especially in Greek), and to recover ancient practices and to practise them anew: these ancient practices were those of the early – the primitive – Christian church.

The Renaissance and the Reformation have one other important feature in common in the way they have been treated by historians. For the Reformation, just like the Renaissance, is usually treated as if it was a purely intellectual phenomenon, related only incidentally or contingently to the particular economic and social circumstances of the place and time where it arose.[33] Yet we have, in fact, every ground for believing that the Reformation was an effect, an expression, a product of a similar high point in economic recovery in the north as the Renaissance was in the south.[34] This high point came a little later than in the south. It brought wealth undreamt of, yet not so much as in the south, and understandably it manifested itself at least in part as a reaction against the south. This is what the Reformation was: a revolt, experienced and expressed in terms of religion, of the newly rich north against the even richer south, of Germany against Italy, of Wittenberg against Rome. As Professor Hugh Trevor-Roper (few people's idea of a left-wing materialist historian) has written of it:

> in the 1520s came the revolt: the revolt of Luther. It was not a revolt within the old, mature economy of Europe: it was a revolt of the 'underdeveloped', 'colonial' areas of northern and central Europe, long taxed, frustrated and exploited (as they felt) to sustain the high civilisation of the Mediterranean and the Rhine.[35]

Thus, in using the category of the Reformation and applying it to the history of sixteenth-century anatomizing, it is important that we should not assume that the Reformation was 'merely' intellectual and mental, that it was simply a matter of *religious doctrines*; Yes, indeed, it was an event which revolved around issues of faith, but faith stirred into life by grand-scale economic, social and political changes at levels the participants could not have been aware of.

In general, therefore, in my treatment here, I shall assume there is no useful explanatory function to be achieved by separating the Renaissance and the Reformation into autonomous categories when discussing sixteenth-century anatomizing. And in now adopting the Reformation as my primary explanatory category I am not jettisoning the 'humanistic' dimensions of the Renaissance; it is just that the Renaissance, as the term is commonly understood and used, is insufficient on its own for an adequate understanding of sixteenth-century anatomizing, given that the subject-matter of that anatomizing was the soul, and the primary purpose of engaging in it was admiration of the Creator and His providence for man.

Thus with respect to the history of anatomy I shall be arguing that in the Christian world the investigation of Nature, as represented by anatomizing, up to and indeed long beyond the sixteenth century, was an inherently Christian enterprise; that there was no way of investigating Nature, or reason to do so, without religious, or better put, pious, motives; that an individual's religious persuasion provided him with the 'spectacles' through which he saw Nature, and thus for what he saw Nature to be. My position is built on a different understanding from that held by most historians of science, of what kind of enterprise sixteenth-century investigation of Nature was: that it was 'Natural Philosophy', rather than 'science' in its modern understanding.

(iv)

So we turn again to the soul, to the soul as expressed through the body.

We have seen the origin of the soul and the concept of the soul in ancient Greece. Plato's writings, especially the *Timaeus*, are a basic source for doctrines on the soul. The soul was the very centre of that self-reflexive and self-improving Greek pursuit, philosophy. It was to care for the soul, and to prepare it for its afterlife, that the leisured Greek philosophers, especially the followers of Plato, engaged in philosophizing. The soul and its passions are what led Plato to pronounce on the anatomy of the human body in the *Timaeus*, finding the natural divisions of the structure of the body which were required to correspond with the nature and requirements of the soul.

The soul of the Greek philosophers, especially the Platonists and neo-Platonists, was incorporated into Christianity, where it was accommodated with a Judaic doctrinal heritage. The Greek philosophical concern with teleology was also adopted into early Christianity – the belief that everything in the universe has a purpose, which it strives to fulfil, a purpose given to it by the intelligent planning of the Creator, God. So teleology became inherent to all proper Christian explanation of Nature, and this made many aspects of Greek philosophy acceptable to Christians. It also, of course, made Greek anatomizing acceptable to Christians too, whether it was anatomizing from the Platonic tradition or from that of Aristotle.

In Christianity only a unitary and immortal soul was adopted, without any proper consideration of how it or its faculties could be partitioned or distributed throughout the body (or, indeed, throughout the universe), as the Greek philosophers had thought and as they had deployed it in their understandings of anatomy. This discrepancy does not, in fact, seem to have been of much concern to Christian theologians and philosophers, and the tri-partite soul of Plato, the unitary distributed soul of Aristotle, and the immortal soul of Christianity, seem to have led concurrent existences. In all three cases, when the human died, it was thought to be because the soul had ceased to function or had departed. Certainly there was concern from time to time about the moment when the immortal soul was, supposedly, implanted by God in the human fetus; in such discussions, it seems to have generally been equated with the rational soul of the Platonic tradition, and to have been located in the head.[36]

As was shown in Part 1, Chapter 2 above, when Greek anatomizing was first revived in the Latin West by Mundinus, it adopted – from Galen – the three-part soul of Plato, and saw the human body as divided appropriately to house and serve this soul. This anatomy, as we saw, was an integral part of Natural Philosophy, the domain of inquiry by learned men into natural things, as conducted primarily at the early universities. And Natural Philosophy was itself a religiously structured area of study, created in part to oppose anti-Catholic heresy, and celebrating the power, wisdom, providence and design of God the Creator. The human body that was seen in anatomizing was the body as designed by God as the high point of His creative acts. Natural Philosophy, while built on classical Greek writings on Nature, especially those of Aristotle, was not itself an ancient Greek area of study, but one created specifically to fill certain needs of Catholicism in crisis. Mundinus in his anatomizing presents the human body as created with supreme skill by the Supreme Maker, and Greek teleology – the 'for the sake of what?' – could be applied to explaining why the parts had the shape, position and function that they did. The Greek neo-Platonic comparison of the microcosm (the body of man) and the macrocosm (the universe) fitted these Christian purposes perfectly too. And Mundinus's book was written for the *anatomical demonstration*; the place and the event where this God-centred knowledge was laid before an admiring audience, as the human body was dissected before their eyes. One of the *doctores extraordinarii* explains the text of Mundinus line by line 'and what he has explained according to text and letter, let him demonstrate by visual testimony, and verify in the cadaver itself'.[37] The body is not being tested against a text, nor is a text being tested against the body. Instead, details of the amazing artifice of the Creator are being read out, and then demonstrated by visual testimony and verified – shown to be true – in the cadaver itself. The audience for this demonstration was not limited to medics, but included important people from the university and the town, and it was repeated ritually year after year before such audiences. Public anatomical dissection was not an occasion for pursuing research, but for demonstrating the body as found. Nor, usually, was it a place for teaching research – or even for teaching dissection (at least until Vesalius). It was, rather, a place for *showing* – exposing to gaze – the high point of God's creation. In this sense it was a religious drama which was being enacted in public dissection. There was no 'contradiction' in this practice between anatomy as 'science' and anatomy as 'religion', because the event was neither of these – but anatomy-as-Natural-Philosophy. The papacy had no need to ban the practice of anatomical dissection and demonstration (though nineteenth-century historians tried to accuse them of having done so, and thus of preventing the progress of science), since the event served the purposes of the Catholic Church and helped keep certain kinds of heresy at bay. It was even perfectly possible for an anatomist to be a friend of popes, as we saw was the case with Berengarius and Columbus.

Throughout the period of anatomizing covered in this book, anatomy continued to be part of Natural Philosophy: Vesalius himself called it 'the chief branch of Natural Philosophy'. There was no splitting off of anatomy from God-centred Natural Philosophy by anatomists bent on secularisation and science, as the traditional story would have it, for there were no secularizing anatomists, only religious-minded ones. The situation is beautifully epitomized by that poetic event at a public anatomical demonstration in 1538 in Lyons, conducted by the physician and famous writer, François Rabelais, which we quoted in the Preface, when the corpse spoke. It was heard to speak by one of the most heretically suspect men of the whole sixteenth century, Etienne Dolet. Yet the message it was heard to give was the message of anatomy as Natural Philosophy: the human body is the high point of God's design in the universe and witness to His providence for His creatures. Fortune had failed to humiliate our corpse by having him hanged as a criminal, the lowest form of human life; for he had revenged himself on Fortune by being privileged to be shown as 'the masterpiece of life's Great Artisan' (that is, of the Christian God). He was not privileged by being the means of conveying important medical knowledge to medical students, but as being shown as an instance of the Creator's art or skill. His role was primarily natural-philosophical, not medical. All anatomy and anatomizing in the sixteenth and surrounding centuries had this as its primary aim.

It is time to attempt to put the religion back into sixteenth century anatomizing. The primary purpose and goal of anatomizing in the sixteenth and surrounding centuries was the knowledge and admiration of God in His creation. The sixteenth century is the century of the Reformation, and all the anatomists we have dealt with and whom we have seen resuscitating anatomical practices of different Ancients, lived and worked in places where discussion of, and conflict over, the new faith was of great consequence. The new faith did not raise issues about the nature of the immortal soul of Christianity – but it most certainly raised pertinent issues about who had the responsibility for its welfare and eternal salvation, the individual or the Church. Moreover, the Lutheran Reformation, the 'Radical Reformation', and the 'Catholic Reformation' each encapsulated certain distinctive attitudes to personal behaviour and forms of personal activity in this world, in order to best prepare the Christian for life in the next. These attitudes were directly linked to views of the nature of God and of the Christian's relation to God in each of these new religious traditions.

If anatomizing gives one views of God, and insights into His attributes, then if one were to change one's view of God and of one's proper relationship to Him, would one need to change one's approach to anatomy? Would one, that is, perhaps feel impelled to engage in *new* anatomizing to cope with and correlate with one's new views of God and of one's proper behaviour as a true Christian? If so, then this might give us a handle to understanding why there was a dynamism to anatomizing in this century of extraordinary religious change, and why particular Modern anatomists turned to emulating particular Ancient anatomists. We turn now to the Modern anatomists in their Reformation contexts.

Notes

1. Wallace K. Ferguson, *The Renaissance in Historical Thought*, The Riverside Press, Cambridge, MA, 1948. See also the article 'Idea of Renaissance' by Denys Hay, in *Dictionary of the History of Ideas*, ed. Philip P. Weiner, (New York, Charles Scribner, 1973), vol. 4, pp. 121–29. Tinsley Helton, ed., *The Renaissance: A Reconsideration of the Theories and Interpretations of the Age*, University of Wisconsin Press, 1964, includes a number of reassessments (mostly built on Ferguson), including Edward Rosen, 'Renaissance Science as Seen by Burckhardt and his Successors', pp. 77–103. For a recent defence of Burckhardt's view, and of the autonomy of intellectual history, see William Kerrigan and Gordon Braden, *The Idea of the Renaissance*, Johns Hopkins University Press, 1991.
2. On the precise political position which led Burckhardt to write the cultural history of the Renaissance, see John R. Hinde, 'The Development of Jacob Burckhardt's Thought', *Journal of the History of Ideas*, 1992, 53, pp. 425–36.
3. From Symonds's article 'The Renaissance' in the 11th edition of the *Encyclopaedia Britannica*, 1910–11, vol. 23, p. 84

4. Quoted in Ferguson, op.cit., p. 200, from Symonds's *Renaissance in Italy*, first published 1875–86; third edition London, 1897, vol. 1, p. 22.
5. This approach is shown at its most extreme in Warren Treadgold, ed., *Renaissances Before the Renaissance: Cultural Revivals of Late Antiquity and the Middle Ages*, Stanford University Press, 1984; Treadgold finds seven renaissances before the Renaissance!
6. Ferguson, op.cit., p. 182.
7. Jacob Burckhardt, *The Civilisation of the Renaissance in Italy: An Essay*, tr. S.G.C. Middlemore, London, Phaidon Press, 1944; 7th printing, 1965, p. 81.
8. Burckhardt, op.cit., pp. 84–85. On Romantic views of genius see Simon Schaffer, 'Genius', in Andrew Cunningham and Nicholas Jardine, eds, *Romanticism and the Sciences*, Cambridge University Press, 1990, pp. 82–98.
9. Ferguson, op.cit., pp. 168–73, 185–88.
10. Burckhardt, op.cit., p. 303.
11. Burckhardt, op.cit., p. 305.
12. Burckhardt, op.cit., p. 107.
13. Ibid.
14. Ferguson, op.cit., p. 170.
15. Erwin Panofsky, *Renaissance and Renascences in Western Art*, London, Paladin, 1970, p. 19. On the construction of the Romantic myth of the Greeks, see E.M. Butler, *The Tyranny of Greece over Germany*, Cambridge University Press, 1935; on the creation of the new domain of 'aesthetics' in the early nineteenth century, see Mary Warnock, *Imagination*, Faber and Faber, 1976.
16. Ferguson, op.cit., p. 177.
17. Ferguson, op.cit., p. 17.
18. On the general issue of economic recovery, see *The Cambridge Economic History of Europe*, vol. IV, 'The Economy of Expanding Europe in the Sixteenth and Seventeenth Centuries', ed. E.E. Rich and C.H. Wilson, Cambridge University Press, 1967, esp. Chapter 1, by Karl F. Helleiner.
19. This is not to imply that the soul dealt with in Christianity (the immortal soul) and the soul dealt with in anatomizing were identical. This question will be dealt with below.
20. The distinguished scholar Paul Kristeller has spent much of his career arguing against certain features of the nineteenth century concept of the Renaissance (for instance, he has argued that *humanitas* did not mean humanitarian/humane; that atheism was rare); see Paul Oskar Kristeller, ed. Michael Mooney, *Renaissance Thought and its Sources*, New York, Columbia University Press, 1979, and his 'The Myth of Renaissance Atheism and the French Tradition of Free Thought', *Journal of the History of Philosophy*, 1968, 6, pp. 233–43. But nevertheless even he is arguing for a revised view of the relation of the Renaissance to religion, not for an understanding of the Renaissance in which religion is central and constitutive.
21. Unfortunately there is no study on the history of the concept of the Scientific Revolution equivalent to that of Ferguson on that of the Renaissance; historians of science have been very unconcerned with the history of their own discipline and its concepts. But see Andrew Cunningham and Perry Williams, 'De-centering the 'Big Picture': *The Origins of Modern*

Science and the Origins of Modern Science', *British Journal for the History of Science*, 1992, **26**, pp. 407–32; Harcourt Brown, 'The Renaissance and Historians of Science', *Studies in the Renaissance*, 1960, 7, pp. 27–42.

22. Burckhardt, op.cit., p. 173.
23. See Nicholas Jardine, 'Legitimation and History', Chapter 6 of his *The Scenes of Inquiry: On the Reality of Questions in the Sciences*, Oxford, Clarendon Press, 1991, esp. p. 129.
24. A. Rupert Hall, *The Revolution in Science, 1500–1750*, London, Longman, 1983, pp. 5–6.
25. General introduction to Marie Boas, *The Scientific Renaissance, 1450–1630*, London, Collins, 1962, pp. 5–7.
26. The current range of opinion can be seen in David C. Lindberg and Ronald L. Numbers, eds, *God and Nature: Historical Essays on the Encounter between Christianity and Science*, University of California Press, 1986; for another example see Brian Vickers ed., *Occult and Scientific Mentalities in the Renaissance*, Cambridge University Press.
27. But see Cunningham and Williams, op.cit.
28. Andrew Wear, 'The Heart and the Blood from Vesalius to Harvey', in *Companion to the History of Modern Science*, ed. R.C. Olby, G.N. Cantor, J.R.R. Christie and M.J.S. Hodge, London, Routledge, 1989, pp. 568–82, see p. 580.
29. For an overview and examples, see Lewis W. Spitz ed., *The Reformation: Basic Interpretations*, Lexington, D.C. Heath and Company, 1962, 2nd. edition, 1972.
30. Symonds, op.cit. (note 2 above), p. 84; he is using the term 'scientifically' here in the sense of 'scholarly'.
31. Quoted in Ferguson, op.cit. (note 1 above), p. 176.
32. For a recent reassessment of the relation of humanism to religious reform, see Peter Matheson, 'Humanism and Reform Movements', in Anthony Goodman and Angus MacKay, *The Impact of Humanism on Western Europe*, London, Longman, 1990, pp. 23–42. On Aléandre, see J. Paquier, *Jérome Aléandre de sa Naissance à la Fin de son Séjour à Brindes (1480–1525)*, Paris, Leboux, 1900.
33. For the controversy on this issue see Spitz., op.cit. (note 29 above). For a recent example, see Alister McGrath, *The Intellectual Origins of the European Reformation*, Oxford, Blackwell, 1987.
34. See *The New Cambridge Modern History*, vol. II, 'The Reformation, 1520–59', ed. G.R. Elton, esp. Chapter 1, 'Economic Change', by F. Luetge; see p. 29.
35. H.R. Trevor-Roper, *Religion, the Reformation and Social Change, and Other Essays*, London, Macmillan, 1967, pp. 32–33.
36. On this general issue see Vivian Nutton, 'The Anatomy of the Soul in Early Renaissance Medicine', in Dunstan, G.R., ed., *The Human Embryo: Aristotle and the Arabic and European Traditions*, University of Exeter Press, 1990, pp. 136–57; and his 'Wittenberg anatomy' in Ole Peter Grell and Andrew Cunningham, eds., *Medicine and the Reformation*, London, Routledge, 1992.
37. Paduan University statute of 1465 or before; see Jerome J. Bylebyl, 'Interpreting the *Fasciculo* Anatomy Scene', *Journal of the History of Medicine*, 1990, **45**, pp. 285–316; see p. 310, and p. 309, note 66.

CHAPTER EIGHT

The Reformation and Anatomising

How might one go about looking at the relation of particular forms of religious commitment to particular forms of anatomizing, especially where one lacks any incontrovertible evidence of religious persuasion on the part of the anatomists in question? In this chapter it will be seen that, on the basis of our present evidence and using our present analytical categories, we can sometimes make strong and conclusive associations between the new kinds of anatomizing brought in by particular anatomists and their local religious contexts and commitments – and thus legitimately conclude that the new kind of anatomizing was the result of the adoption of one or other new kind of religious confession. In some cases the correlation is weaker but suggestive and could perhaps be legitimately inferred; while in other cases it would still almost require the eye of faith itself to see it. On the principle that actions speak at least as loudly as words, one approach I shall use here, in what is avowedly a tentative essay, is to seek a homology of actions: seeing what parallels and similarities one can find in the actions of religious reformers on the one hand, and the actions of contemporaneous Modern anatomists on the other, and to see what inference (if any) one can reasonably make about religious commitment on that basis, and whether a commitment to a particular new or old religious confession underlies a particular kind of anatomizing. Even when our correlations turn out to be on the meagre side, in the following case studies it will be necessary to deal quite extensively with some of the major events of the Reformation in order to locate our anatomists with respect to them.

(i) Erasmian Reform

Our first anatomists of the Renaissance, were Sylvius and Guenther, working together at Paris in the 1520s and early 1530s trying to restore the true doctrine of Galen. They were contemporaries of Erasmus (*c.* 1469–1536), the most famous of all literary humanists and the man said to have laid the egg that Luther hatched and hence inadvertently starting the Reformation. In this case there are suggestive parallels which can be made between the activities of these reforming anatomists and this great religious reformer, but there is no possibility of making any firm link between religious confession and attitudes to anatomizing.

Sylvius and Guenther were trying to restore the true doctrines of Galen, by turning back to the Greek text of Galen's great anatomical works, editing them, translating them, reducing them to compendia, and teaching as widely as possible this new truth about Galen and the anatomy of the human body. Their views, as we saw, were not radical: they were not prepared to find fault with Galen, or to go beyond him to yet earlier anatomists. The views of Sylvius and Guenther with respect to Galen's anatomy, and their activities in restoring and reviving Galen's doctrines in their pristine purity, resemble so closely the work of Desiderius Erasmus with respect to the text of the Bible that they might have been patterned on it. Erasmus called on his contemporaries: 'Let us read the Gospel with ardent efforts, but let us not only read it but also live it.' Perhaps Sylvius and Guenther might be metaphorically portrayed 'living the Gospel' similarly in anatomy.

Erasmus came from Rotterdam, in the Netherlands. This area, though under Hapsburg rule, was the closest thing to a northern rival to the city-states of Italy, and Antwerp was in Erasmus's day becoming the great commercial centre of the north. He was among the first northerners to acquire the humanist linguistic skills: Ciceronian Latin, and a mastery of Greek. He dedicated these skills to translating, editing and publishing many of the newly discovered works of ancient Rome and Greece, and hence helped popularize them. And he was also just about the first person who applied these humanist skills to the study and revision of the biblical texts; it is in doing this that he is said to have laid the egg.

The illegitimate son of a priest, Erasmus was educated by the Brethren of the Common Life, a somewhat mystical order, and then, with great reluctance, joined the Augustinian canons and became both a monk and a priest. It was when he became Latin secretary to the Bishop of Cambrai in 1494, however, that he was able to exercise more freely his enthusiasm for 'good letters' (*bonae literae*), and especially the Latin poets. The bishop gave him permission to pursue a theology degree in Paris, the first of many expedients Erasmus tried to ensure that he would never again have to return to the closed life of the monastery. While in Paris Erasmus published his first famous collection of 'Adages' from classical writers, aimed at increasing people's familiarity with good Latin writing.

A turning-point came on Erasmus's visit to England in 1499 where he heard John Colet lecturing at Oxford on the Epistles of St Paul. Colet was among a group of devout Englishmen who had come under the influence of the work of Ficino's circle in Italy. As a consequence, Colet was trying to explore what might be called the 'mystery' or spiritual dimension of St Paul's writings via certain ancient neo-Platonists, especially Dionysius the Areopagite. Colet's spiritual journey has been brilliantly decribed by Frederic Seebohm, who has traced the way in which Colet had 'risen out of Neo-platonism to Dionysius, and from Dionysius to St Paul', and beyond, to Christ.[1] Inspired by this new approach Erasmus, the priest and classical enthusiast, realized at the relatively late age of about 30 that he had to learn Greek: 'Latin erudition is crippled and imperfect without Greek', as he now wrote.[2] Erasmus's publications, such as the *Enchiridion of the Christian soldier* (1503), now explicitly discuss the importance of classical study, and especially Plato, as a means to an end; and that end is the cultivation of the inwardness, the inner experience of spiritual life (as opposed to liturgy and ritual). Erasmus was now turning his linguistic skills and classical learning to developing a *Christian* journey of the spirit through to the highest mysteries. Eventually he was able to make a kind of pilgrimage to the centre of all these studies, Italy (1506–09), and there he got caught up with the 'Academy' of the printer Aldus Manutius in Venice. In the next few years he turned his increasing knowledge of Greek toward translating the Greek Fathers and, most significantly, toward producing the first-ever printed edition of the Greek text of the New Testament and a new Latin translation of it (1516).

When he produced his new New Testament in Latin, Erasmus assured his reader that his intention was not to replace the version used by the Roman church, the Vulgate, but to make it 'more correct, and clearer' (*emendiator, magisque perspicua*), and hence he had compared the traditional text with Greek and other Latin manuscripts. 'Let us therefore read the Gospel with ardent efforts, but let us not only read it but also live it' (*legamus igitur ardentibus studiis evangelium, sed non tantum legamus, verum etiam vivamus*).[3]

What Erasmus meant by 'living the Gospel' becomes more clear from the famous 'Paraclesis' that he put as a preface to his translation. The philosophy of Christ, he argues, is superior to all other philosophies; it promises the highest wisdom and mysteries; only by reading its basic texts (the Gospels) will we master it.

> Only this Master and Teacher came from heaven, which alone could teach sure things being the everlasting wisdom of the Father . . . To lead an angel life, that is the office and duty of a Christian and divine . . . If schoolmasters would instruct their children rather with this simple science than with the witty traditions of Aristotle and Averroes, then should the Christianity be more at quietness, and not be disturbed with such perpetual storms of dissension and war . . . We cannot call any man a Platonist unless he have read the works of Plato. Yet we call them Christian, yea and divines, which never have read the scripture of Christ. Christ sayeth, he that loveth me doth keep my sayings . . . The first point of Christianity

> is to know what Christ hath taught. The next is to do thereafter and to fulfil it as nigh as God giveth us grace.[4]

To be a proper Christian is to follow the sayings of Christ; we can discover these sayings, this wisdom, directly from the Bible; if we all were good Christians, we would be at peace with each other and thus the world would be at peace. Christ's is the highest philosophy, and it leads to peace and concord.

Thus it was a northerner, Erasmus, who turned the new learning to the task of retranslating the Bible, and making it more widely available. Because of his own humanist sympathies, Erasmus saw the Bible primarily as a book of philosophy, of wisdom, of the highest mysteries. Plato had seemed to promise 'pagan mysteries'; Erasmus (in the words of a recent scholar)[5] busied himself in 'christening' those pagan mysteries. And other northerners shared his interest: Englishmen like Colet, Grocyn and More, Frenchmen like Jacques Lefèvre d'Etaples. This south – north difference can perhaps best be seen in the fact that in 1513 the Greek text of Plato's complete works, prepared by refugee Greeks in north Italy and coordinated and printed by Aldus Manutius in Venice, was dedicated to the Medici pope, Leo X, whose education had been in the care of Marsiglio Ficino and Pico della Mirandola. The dedication by Aldus claims that 'the works of this divine man', meaning Plato, 'could be dedicated to no-one more appropriately than to you, the highest priest', that is the pope, 'of divine matters', or theology.[6] That was a southern production. Only three years later it was to this same pope that Erasmus, the northerner, dedicated his new translation of part of the *Christian* divine text, the New Testament, printed beyond the Alps, in Basel.

This is a rough sketch of a famous humanist, a reformer but not a revolutionary, a worker for peace and concord who believed that the translation of central texts would work toward common doctrine and peace. Sylvius and Guenther, contemporaries of Erasmus working in anatomy, were humanists who had been educated in Erasmus's world, and had possibly even learnt their Greek and Ciceronian Latin from Erasmus's own books. Do the parallels go further, and do they allow us to infer anything about the religious position of these anatomists? And allow us to infer any inspiration from their religious views to their work on anatomy?

All three men came from the same generation, and they all had similar educational experiences. They were all part of that seizing of the initiative in intellectual matters by men of the north which for so long had been typical of the south, and especially Italy. In mid-life they all felt that 'Latin erudition is crippled and imperfect without Greek'. They each used their newly acquired knowledge of Greek to make available in Latin the fundamental text of their respective fields: the New Testament, and the anatomical writings of Galen. Thus they could make their respective favoured texts 'more correct, and clearer', and restore the 'primitive', pure and pristine doctrine of their respective great ancient authority. They exhorted their audiences to discover what the authority had said, and to follow it as doctrine. As Erasmus wrote,

> Christ sayeth, he that loveth me doth keep my sayings . . . The first point of Christianity is to know what Christ hath taught. The next is to do thereafter and to fulfil it as nigh as God giveth us grace.

'The knowledge of anatomy,' writes Guenther in comparable voice, 'useful and necessary not only to physicians and philosophers but also beautiful and honourable to all kinds of men, today lies utterly neglected.' In recovering this essential knowledge, Guenther leans totally on Galen: 'As for my contribution,' Guenther writes, 'there is nothing written here which is not redolent of Galen, whose doctrine I deliberately follow: there is nothing asserted which I have not seen him write.'

The reforming activities on both sides certainly made it possible for Erasmus to have a new vision of how Christianity should be renewed, and for Sylvius and Guenther to have a renewed view of the Galenic body. Whether the two missions and visions were prompted by the same religious feelings is, however, not possible for us to ascertain, even though Guenther in later life was to convert to a mild form of Protestantism. The parallels in their respective reforming campaigns are interesting but inconclusive.

Both Erasmus and Sylvius were to fervently maintain their non-radical reading of their ancient texts when challenged by someone in the next generation who was prepared to go much further: Luther and Vesalius respectively. Possible links between religious commitment and anatomizing programmes become clearer in the next generation.

(ii) Lutheran Reformation: The Word

Luther, it was said, hatched the egg that Erasmus had laid; in other words, he took seriously the implications of the kind of reform that Erasmus had called for – with momentous effects for the whole of Europe. It might be said that Vesalius promoted a parallel revolution in anatomy: He taught the world to see a different body; he taught anatomists, physicians and philosophers to adopt a new ambition with respect to the Ancients. It is not particularly novel to call Vesalius a 'Luther of anatomy'. What has not been done before, however, is to look and see the ways in which his anatomical enterprise and achievement were in their own time 'Lutheran'. Here the inferential evidence is significantly stronger than in the previous case, though perhaps not totally conclusive.

Martin Luther was a troubled Augustinian monk and professor of theology tucked away in the new little university of Wittenberg, and he had been wrestling with the implications of the fact, central to the teaching of St Augustine, of man's fallen nature.[7] Luther took this to its logical conclusion: if man is fallen, then he is in bondage to sin and nothing he can do can justify him in the eyes of God. His sinful nature prevents man from following the law of God; yet man cannot be saved by his own efforts. Luther's solution to this, the 'blinding flash' that happened to him in the tower room in 1513 as he read the Psalms and Romans, was that justification of man was indeed possible, but only by trusting in God's 'righteousness' to save. That is to say, each individual is the potential recipient of God's abounding grace and forgiveness. All that each individual has to do is to actively accept this grace, an acceptance which can be made 'by faith alone', and then he will be 'reborn' in Christ. 'Good works' are powerless and irrelevant to salvation. Luther had found in Augustine what looked to him like a problem, and solved it through a particular reading of the Bible.

In this way of looking at the relationship of God and man, the only mediator necessary for salvation was Jesus Christ, who was himself the Son of God. Suddenly there was, for Luther, no role at all for the mediation of a special priesthood or for the apparatus of the Church, for they were presenting themselves as the essential intermediaries in a relationship which demanded that there be no intermediaries. Thus the whole Church organization, from the pope down to the parish priest, was redundant as a means to the salvation of the individual. And what is more, Luther could find no biblical warrant for the existence of a Church in the Roman Catholic form, nor for many of its practices, such as indulgences, the veneration of saints, the ritual use of relics. In his famous 95 theses of 1517 Luther challenged the Church authorities to justify in scholastic disputation – to show their textual warrant for – the practice of indulgences. Appropriately Luther pinned this challenge to dispute to the university noticeboard, which was the door of the university church of Wittenberg.

In making this resolution of his personal spiritual problems, what Luther thought and claimed he was doing was bringing back into existence the *primitive apostolic church*: the congregation of all believers. He was going back to the early church, the one that Christ intended; going back, that is, before the institution and (as he saw it) corruption of the Catholic Church. The real church was not a formally organized hierarchy of power, but one body of men, united spiritually, each with the responsibility of seeking his own salvation. To bring about this state of affairs Luther, like Erasmus, advocated the education of each individual in the articles of faith (catechizing). But Luther also insisted on *the duty of every person to read God's Word (the Bible) for himself*. Erasmus had translated the New Testament anew into what was by now the language of the learned, Latin; but Luther translated it into the language of the people, German. He published his New Testament in 1522, the complete Bible in 1534. And he saw to it that the university of Wittenberg was henceforth devoted to the creation and propagation of what was in his eyes a properly Christian curriculum of studies. Under the direction of Luther's soul-mate Philipp Melanchthon, 'the preceptor of Germany', it became the mother of Protestant universities.[8] Thus Christianity was, for Luther, a religion of personal engagement; it is the Word of God which is (quite literally) the guide to salvation and the true Christian life, and every individual needs to read and hear it for himself. The Word, in the sense of the text of the Bible, was of central importance to him. In his confrontation with the defenders of the Catholic Church, Luther's challenge was always for them to show – to literally point out with their finger – the texts in the Bible on which their claims were built. Only this would count as authority to Luther. If they could not point to a pertinent text, their claims had no authority in Luther's eyes.

Luther and his followers had a new goal: to live the life of true apostolic Christians. The Word, the Sacrament (a reduced and redefined set of rituals), and Discipline, these were to be the keys to the new Christianity – which was, in Luther's view, the oldest form of Christianity. Catholics and Protestants now looked to the same book, the Bible, but with differing interpretations of what it taught. For the Protestants its message was that they should *be* primitive Christians, like the early Apostles.

Some of these features of the doctrine of the reformed faith can be seen dramatically in the iconography of Luther, that is, in the pictures painted of Luther and under the influence and impact of his doctrines. On this theme we shall follow Carl Christensen's excellent recent analysis.[9] Almost alone among the reformers, it appears, Luther remained favourable to the use of imagery in church. Some of the reformers of the Swiss cities, for instance, were to pursue a drastic policy of clearing out and breaking all statues, pictures and relics from their local churches, on the grounds that the worship of 'idols' was forbidden by the Commandments. Their fear was that worship of what the image represented actually became worship of the image itself, and hence idolatry. Luther, however, held the view that as we unavoidably form mental images of things that we hear described in words, so it is positively desirable to look at a pictured image which properly represents the 'thing' in question. Hence for Luther pictorial images had a useful pedagogical role in spreading the new doctrines. They are also good propaganda. Fortunately Luther's protector, the Elector Frederick the Wise of Saxony, had installed a court painter of some talent, Lucas Cranach the Elder (1472–1553), who was succeeded by his son, the Younger (1515–1586). Between them the Cranachs produced many portraits of the Wittenberg reformers and, from 1529 onwards, a number of pictures which portray several of Luther's main doctrines. So keen was Luther on the pedagogic value of illustration that he had the elder Cranach produce a number of woodcuts for his German translation of the Bible of 1522: Luther himself decided which themes should be portrayed and, to some extent, how.[10]

A favourite theme of both Luther and the Cranachs was the Last Supper (a relatively rare theme of religious art hitherto). As Luther wrote in 1530:

> Whoever is inclined to put pictures on the altar ought to have the Lord's Supper of Christ painted, with these two verses written around it in golden letters: 'The gracious and merciful Lord has instituted a remembrance *of His wonderful works*'. Then they would stand before our eyes for our heart to contemplate them, and even our eyes, in reading, would have to thank and praise God. Since the altar is designated for the administration of the Sacrament, one could not find a better painting for it.[11]

If we look at the altar-piece by the younger Cranach which stands today in the church at Dessau, we can see an extraordinary portrayal of this theme (see Figure 8.1). For here we have Christ surrounded by the Apostles; and those 'apostles' are Luther, Melanchthon and other leaders of the Wittenberg reformation! The reformers are shown as *being* the early Apostles. And of course there is no priest acting as intermediary. Christ himself is distributing the sacramental elements directly to these apostles. What is more, communion 'in both kinds' (that is, bread and wine) is being taken, and this is another apostolic practice that Luther worked to reinstitute. Another altar-piece by the elder Cranach (see Figure 8.2), still today in the church at Wittenberg, has a comparable portrayal of Luther and others as the disciples at the Last Supper. And on the wings of this there are pictures which further show the reformers as engaged in revived apostolic practices; the unordained Phillip Melanchthon as St John the Baptist, performing infant baptism; and the pastor of the church (one Johannes Bugenhagen), as St Peter holding the keys of the Church! The message about how the reformers were living out primitive apostolic practices could hardly be more direct.[12]

It is important to appreciate the exclusiveness of the source of authority to which Luther was appealing. The practices, beliefs and behaviour

8.1 Luther and his followers as the disciples of Christ at the Last Supper. Lucas Cranach the Younger, Dessau Altar-piece, 1565, Schlosskirche, Dessau.

of proper Christians are to be discovered only from Scripture. Scripture – written Scripture, the text of the Bible – is the arbiter, the ultimate authority. For Luther an issue is resolved when he can point to the relevant words in Scripture. This is not just a simple-minded literalism on his part. For the Word had a very special status for him. Hear him ecstatic in *The Freedom of a Christian*:

8.2 Wing panels of the altar-piece by Lucas Cranach the Elder, 1547, 'Memorial of the Reformation', now in Wittenberg Stadtkirche, showing Luther's lay companions taking up what had been priestly roles.

> One thing, and only one thing, is necessary for Christian life, righteousness and freedom. That one thing is the most holy Word of God, the gospel of Christ . . . 'Men shall not live by bread alone, but by every word that proceeds from the mouth of God'. Let us then consider it certain and firmly established that the soul can do without anything except the Word of God and that where the Word of God is missing there is no help at all for the soul. If it has the Word of God it is rich and lacks nothing since it is the Word of life, truth, light, peace, righteousness, salvation, joy, liberty, wisdom, power, grace, glory, and of every incalculable blessing. That is why the prophet in the entire Psalm [Psalm 19] and in many other places yearns and sighs for the Word of God and uses so many names to describe it . . . On the other hand, there is no more terrible disaster with which the wrath of God can afflict men than a famine of the hearing of his Word . . .[13]

The Word is, for Luther, three complementary things: it is the Scriptures, it is Christ, it is God's Law. It is all these at once – and this is the source of its authority for him. Hearing the Word, preaching the Word, reading the Word; these are central activities for a Lutheran Christian to engage in. And nothing was more powerful for Luther than the promises from the mouth of God which he could find in the written Word of God; his whole theology is based on the fulfilment of God's spoken promises. The Word was very powerful. It accomplished everything. In one of the sermons he preached against the enthusiasts at Wittenberg in March 1522, Luther ascribed to the Word alone all his success at the Diet of Worms:

> I opposed indulgences and all the papists, but never with force. I simply taught, preached, and read God's Word. Otherwise I did nothing. And while I slept, or drank Wittenberg beer with my friends Philip [Melanchthon] and [Nicholas von] Amsdorf, the Word so greatly weakened the papacy that no prince or emperor ever inflicted such losses upon it. I did nothing; the Word did everything.[14]

Luther quickly gained many adherents. The source of Luther's own personal problems, to which his reform of religious doctrine and organization was his solution, has been very persuasively located by a psychoanalytical writer, Erik Erikson, in the place one would expect him to locate it: in Luther's own youthful experiences.[15] The son was rebelling against his father: Luther the man against his human father, Luther the monk against the pope, his spiritual father. But why should the appeal of the doctrines of Luther have been so great to people who had not shared Luther's own personal dilemmas? Luther's was far from being the first heresy within the Catholic Church: why was it the most successful, both in itself and in the more extreme forms (such as Calvinism) that it spawned?

One of the positive appeals of Luther's new doctrine was that it was, in its rituals, a religion of participation. This is particularly evident in Luther's teaching on the ritual of communion. Further, each person's conscience, regulated and informed by the Word of God, was to be the final arbiter of right and wrong, not the volumes of canon law – which Luther ceremonially burnt in 1520 in response to his own books being burnt by the pope. To potential converts, therefore, 'Lutheranism' may be said to have meant a new stress on the responsibility of the individual for the fate of his own soul; a new stress, that is, on individual autonomy.

When Luther 'went public', with the 95 theses, there was great consternation in Catholic circles. Luther was ordered to be silent, but he refused.[16] The theses were refuted in print by Johannes von Eck of Ingolstadt, defended, counter-refuted, re-defended all within the space of eight months. Then Eck challenged Luther to a public disputation, which took place at

Leipzig in mid-1519. The unwillingness of Luther to disown his views at the disputation obliged Eck in 1520 to get a papal Bull excommunicating him. Seen from the perspective of the pope, Luther's message was a heresy threatening the status quo with immense social and political dangers. The emperor's support was courted by Luther, but soon the emperor too came to see the reform as a threat to social and religious order. Pope and Emperor were quickly proved right.

Now we turn to comparing the activities and claims of Luther with those of Vesalius the anatomist. We will be looking to see whether or not Vesalius undertakes new anatomizing initiatives and sees a new body *because* he has adopted Lutheran attitudes. This of course will involve establishing that Vesalius adopted Lutheran attitudes. If the comparison is successful, then it might be possible to say that just as Luther went beyond Erasmus, and followed in his own practice his understanding of what primitive Christians *did* as well as follow what they *said*, so Vesalius similarly went beyond Sylvius and Guenther and in his anatomizing followed what Galen *did* as well as what he *said* – and went even beyond that. The first thing we need to establish is whether Vesalius came into contact with reformed religious teaching.

The appeal of the reformed doctrine was primarily in the north of Europe. The most fundamental reason for this seems to be the great shift of economic power that was taking place at this time; a shift from the Mediterranean to the north of Europe, from Italy to the Netherlands, from Venice to Antwerp and Amsterdam. Though it would take another century to complete, the social and political effects of this economic shift were already visible. It was marked by the renewed growth of cities as centres of wealth in the north. The Lutheran reformation was to be primarily an urban phenomenon, and one could say that with it the city people of the north put their confession where their interest was.

Vesalius was from the north. Charles of Ghent (1500–1566), was in 1519 unanimously elected as the Holy Roman Emperor, and became Charles V. While Vesalius was an infant, his father was with the Imperial court, which at that time was in Spain (though young Vesalius may have been left at home). Educated in the Netherlands himself, the young emperor had taken with him to Spain courtiers who were admirers of Erasmus of Rotterdam. During these years, even though the Inquisition was active in Spain, the cultivation of Erasmian ideals enjoyed an unparalleled vogue, especially at court. Charles V, though he became the hammer of the Protestants, carried his private Erasmianism with him to his grave.[17] Vesalius's father was a member of this court.

The exciting events of Luther's rebellion against Rome were happening during Vesalius's youth in the Netherlands, and in an area relatively close by. It spilled over into the territories in which Vesalius lived.[18] Indeed the urgencies of the situation created by Luther and his followers intermittently recalled Charles to the northern parts of his empire for the rest of his reign.

After his time at the new humanist tri-lingual college at Louvain (1530–), where he learnt Greek and Hebrew and the best humanist Latin, Vesalius could have read the Bible in the original languages if he had wanted to. So from all this, we know that by the time he was 19 years old, Vesalius would have been aware of, and perhaps even quite familiar with, Erasmian thinking about religious reform.

In 1533 Vesalius went to Paris to study medicine. A striking feature of Paris's university at this time was that it was a bastion of religious orthodoxy. If Vesalius's parents had wanted a place for their son to learn medicine which was most likely to discourage him from the new heresy, Paris would at one time have looked ideal. For it was there that Luther's doctrines had been most thoroughly condemned in 1521. But now in 1533 the impact of Lutheran teaching was once again impossible to ignore even at Paris. The Sorbonne and the Parlement of Paris were busily hunting out and condemning heretics: Louis Berquin had been burned alive for heresy as recently as 1529. Indeed the Syndic of the Sorbonne, Noel Beda (Bédier), seriously regarded the learning of Greek as a potentially heretical activity, summonsing the Royal Lecturers before Parlement in 1534.[19] Among the generation of Vesalius's fellow students at Paris were Jean Calvin in the law faculty and Michael Servetus in medicine (of whom we shall be hearing more anon). In November 1533 the student rector of the university, Nicholas Cop, son of the famous physician, made a notorious speech (reportedly written by Calvin) which led to both him and Calvin having to flee France for their lives. Then in October 1534 there was the 'Affair of the Placards', the outrageous attack on the mass. To have failed to notice the effects at Paris of Luther's claims, Vesalius would have had to have been a hermit; we can, I think, take it that he was familiar with them.

Was Vesalius a Lutheran, a committed follower of this new primitive Christianity? And was such a commitment on his part the source, spring and reason for his reform of anatomical practice? Did his anatomizing stem from the Reformation proper through the person of Vesalius the follower of Luther?

In the first place we have to consider the argument from silence, for there is no mention in writing by Vesalius himself, or by any of his contemporaries, of his religious persuasion. To be Catholic was to be unremarkable, and no one remarked that Vesalius was not Catholic. Secondly, after Vesalius had finished his great book, the *Fabrica*, he went off and became a physician to the emperor, Charles V, refusing all offers of university posts.[20] Charles spent his life trying to cope with the mess that Luther had created, trying to bring erring parts of his empire back to the true faith, trying to keep his empire united under one faith, Catholicism. Vesalius survived some dozen years in this court, close to the person of the Emperor. It would have been quite a feat for Vesalius, who as we have seen was highly opinionated and argumentative – 'choleric', as the student note-taker at Bologna called him – to keep quiet about his aberrant beliefs, if he had such beliefs. The argument from silence thus indicates that Vesalius was not a Lutheran, and indeed that he may have kept his anatomizing separate from his religious commitment, whatever it was.

On the other hand, however, there are Vesalius's actions: the pattern of what he did in his anatomizing, how he did it, and what he called on other people to do, in turn, in their anatomizing.

From his early days in Padua as an anatomical demonstrator, Vesalius, as we saw earlier, constantly placed the 'evidence of the body', acquired by personal experience of dissection or by observation of the dissected body, over the evidence of texts. As he wrote about the azygos vein in the bloodletting letter of 1539: 'No-one who has either engaged in the task of dissecting personally or who has observed a dissection will allege that [this vein] arises under the base of the heart or before the vena cava reaches the heart'.[21] It was argued above, in Chapter 4, that Vesalius could make such claims because he had an unrivalled experience of and skill at dissection, greater than that of his teachers, and also because he had a strong, not to say arrogant, personality. But obviously this is not enough of an answer, for it does not account for *why* he should put the evidence of the body before the evidence of the texts that his teachers revered. It explains only how he had the technical and personality resources to maintain his position, not why he took that position. Taking that position was not easy and direct: the body did not announce itself to Vesalius as a text truer than Galen. 'I read Galen's book *On the bones* to the students at least three times,' Vesalius later wrote, 'before I dared perceive any mistake on Galen's part, although now I can't be sufficiently astonished at my own stupidity at understanding so little what was written, and thus myself deceiving my own eyes.'[22] There had to be some impetus *outside* the body being dissected, outside the experience of dissecting, to enable Vesalius to understand 'what was written' (*quae scribebantur*). And when he did so, when he realized that Galen had made mistakes, he was aware of what kind of act it was to criticize Galen. In the bloodletting letter of 1540 Vesalius talked of criticism of Galen as being potentially 'heretical' and 'impious'; his language in this respect is quite striking, and he raises the issue several times.[23] Although it might sound heretical to say so (he is saying), the body should be the text, and we should satisfy ourselves about what it says by dissecting personally or observing at a dissection.

With Luther, the extraordinarily explicit and repeated anthropomorphizing of 'the Word' that he went in for certainly indicates how strongly he leant on the text of Scripture. This text was, for him, alive, and not only an active agent in his struggles, working on his behalf in the cause of truth, but also the sole authority to which appeal could be made in any dispute. For Vesalius the dissected human body has this same status, as we have seen, and he repeatedly refers to it as a 'text', which can literally be read, calling it 'this true book of ours, the human body, man himself'. No other text has a remotely comparable status, not even the texts of the immortal Galen whom his teachers venerated and whose superior truth Curtius defended in that celebrated 22nd demonstration at Bologna.

This 22nd demonstration is so revealing of Vesalius's views that we should look at it again:

> When the lecture of Curtius was finished, Vesalius, who had been present and heard the refutation of his own arguments, asked Curtius to accompany him to the anatomy [the 22nd of the series]. For he wanted to show him that his own opinion was most true. Therefore he brought Curtius to our two subjects for dissection.
>
> Now, he said, excellentissime Domine, here we have our subjects: we shall see whether I have erred. These we now want to examine; and we want to put Galen aside for a while, for I acknowledge that I have said – if it is permissible to say so – that here Galen has erred because he did not know the site of this 'vein without pair' in the human body, something we have today the same as he did.
>
> Curtius answered by smiling, for Vesalius, as he was choleric, was very excited: No Domine, he said, Galen should not be put aside by us, since he always well understood everything, and consequently we also follow him. Do you know how to interpret Hippocrates better than Galen knew?
>
> Vesalius answered: I am not saying that, but I am showing you here in these two subjects the 'vein without pair', how it nourishes all the lower ribs, except the two upper ones in which pleurisy does not occur. For always here – here he knocked with his hands against the middle of the chest – occurs inflammation and pleurisy, not in the two upper ribs . . .
>
> Curtius replied: I am no *anatomista*, but there can well be still other veins nourishing the ribs and the muscles besides these.
>
> Where, I ask? Vesalius said, Show them to me.
>
> Curtius said: Do you want to deny the ducts of Nature (*meatus naturae*)?
>
> Oh! Vesalius said, you want to talk about the invisible and the occult – but I am talking about the visible (*de manifestis*).
>
> Curtius answered: Indeed I always deal with the most evident (*de apertissimis*). Domine, you do not understand Hippocrates and Galen well on these matters.

> Vesalius replied: It is quite true, because I am not such an old man as you are, etc. . . .
>
> Vesalius said: Domine Doctor, I beg Your Excellency not to think me so unskilled that I do not know and understand these matters.
>
> Smiling, Curtius said: Domine I did not say so, for I have always said that you are excellent, but I have rejected your false exposition of Hippocrates implying that Galen should have erred in these matters.
>
> Vesalius replied: I acknowledge that I have said that Galen has erred in these matters, and this is evident here in these subjects, as also many more mistakes of his.[24]

Curtius had used words and arguments, but Vesalius constantly returns to the body, to looking, showing and pointing. Only the true text, the body itself, can speak with authority, and it is the authority to be preferred in all cases over ancient written texts, whether by Galen or any other venerable Ancient. 'Show them to me' is Vesalius's challenge to Curtius, 'Galen has erred in these matters, and this is *evident here* in these subjects' (my emphasis). Vesalius is rejecting *patres istos* – those Fathers – Galen and Curtius; like Luther rejecting the authority of the pope, Vesalius the young man is rebelling against his fathers: 'It is quite true,' he scoffs at Curtius, I do not understand Hippocrates and Galen the way you do 'because I am not such an old man as you are'.

Not only does Vesalius insist on the primacy of 'the Word', that is, the body, over written text and tradition but, like Luther with the Bible, he introduces touching and pointing, into both the practice of public anatomizing and its visual representation, as aids to witnessing the truth for oneself. In the last of the Bologna demonstrations, Vesalius showed the humours of the eye. 'All these,' he said, 'anybody can see for himself at home; for surely, Domini, he said, you can learn only a little from a mere demonstration if you yourselves have not taken the objects (*subjecta*) into your own hands.'[25] The students did just this; came up and touched the parts for themselves, took them into their own hands. And in the vivisection demonstration, when pressed by the students for an answer as to the relation of the movement of the heart to the movement of the arteries, Vesalius answered: 'I do not want to give my opinion, you yourselves should feel with your own hands, and trust them.'[26] Moreover, on the title-page of the *Fabrica*, Vesalius has the students portrayed not only as close to the body, so that they can *see* it, but as actually reaching in so that they can *touch* it (see Figure 8.3).

So the question recurs: *why* did Vesalius reject one authority, one text (Galen) for another (the body), why did he stress the indispensability of *personal engagement* in dissection, and the crucial importance of seeing for oneself and touching for oneself? In Chapter 4 I argued that in doing these things he was simply following Galen, bringing Galen back

8.3 The students reaching forward. Title-page of the *Fabrica*, 1543, detail (enhanced).

8.4 Vesalius portraying himself as a modern Ancient. Title-page of the *Fabrica*, 1543, detail.

to life. And the iconography of the title-page of the *Fabrica* again portrays this message: Vesalius presents himself *with* the Ancients, engaged in an ancient practice of anatomizing (Figure 8.4). It is exactly equivalent to Luther having himself presented as one of the true primitive Christians, as a disciple of Christ, in the Protestant altar-pieces painted by the Cranachs. But, again, this does not explain *why* Vesalius turned to emulating Galen, why he portrayed himself as a primitive anatomist. That impetus, I suggest, lay outside the world of anatomizing. It is in this act of trying to live out the experience of *being* Galen the anatomist that we might now be justified in seeing Vesalius as a Lutheran – or at the very least as acting here *like* a Lutheran. For this is what Luther did; he tried not just to follow the advice of Christ about how to live properly, but also to *be* a disciple of Christ in the present and to thus restore primitive Christianity.

There are some circumstances which add a little supplementary evidence to suggest that Vesalius may have been a Lutheran. There is the strange circumstance of Vesalius publishing the *Fabrica* in Basle, when Venice was so much nearer. Basle was now so reformist, so Protestant, that even Erasmus had had to leave it for a while. Yet Vesalius did not hesitate to send all the wood-blocks over the Alps in order to have them printed there.[27] Then there is the matter of the circumstances of Vesalius's death. He died returning from pilgrimage to Palestine. Although this could have been a quite innocent and pious journey, pilgrimages to the Holy Land were also (as Cushing has pointed out) one of the penances meted out by the Inquisition to wanderers from the path of truth.[28]

The connection between Vesalian anatomizing and Lutheran Protestantism is strikingly strong. Anatomy – indeed, Vesalius's anatomy – became a Lutheran Protestant endeavour, not in the medical schools but in the far larger arena of the philosophy course of universities, the basic course taken by all students at university. Anatomy was introduced in the years around the preparation of the *Fabrica* and by no less a figure than Melanchthon, Luther's soul-mate and the man who became the 'Preceptor of Germany' through his reforms of the curriculum at Wittenberg, and hence of all the universities brought into existence or reformed through the Lutheran reformation. Melanchthon first became concerned with teaching anatomy to arts and philosophy students at Wittenberg in 1535;[29] he felt it was necessary to restore teaching of the soul to the curriculum, and for this purpose brought back from disgrace Aristotle's *De Anima*, On the soul. He first got his colleague, Milich, to comment on it as part of the physics (Natural Philosophy) course, and then in the second half of 1539 he wrote a commentary on it himself, publishing this in 1540 as *Commentarius de Anima*. Aristotle's *De Anima* had been an important text in the philosophy course for hundreds of years, dealing as it does with the nature and essence of the soul and with its attributes and properties. But Melanchthon was the first person for centuries, possibly the first since Aristotle himself, to treat the *De Anima* as requiring or necessitating a knowledge of anatomy. For what Melanchthon claims is that the *De Anima* cannot be rightly understood without at least some knowledge of the body itself.[30] Melanchthon's reasons for re-introducing Aristotle's discussion of the soul and its operations (with its account of how the soul receives its information from the senses, not from inspiration) arose from a local and urgent threat to the Lutheran religious reforms,[31] and it was thus the immediate political concerns of Lutheran Protestantism which brought anatomy into the philosophy curriculum.

That anatomy was totally God-centred. The knowledge of anatomy, Melanchthon writes (attributing the sentiment to Galen) leads us to the knowledge of God: *Galenus praeclare dixit, Anatomiae scientiam, ducem nobis esse ad Dei cognitionem.*[32] Melanchthon gives a quite extensive listing of the outer parts of the human body, and then turns to the inner ones:

> At this point truly you think yourself introduced into a temple and a kind of shrine; on this account you ought not to simply look at the material with particular reverence, but to take into consideration the plan and diligence of the Maker. For the scheme of the work bears witness that men do not exist by chance, but take their rise from some infinite Mind which has arranged its individual parts with astonishing care and destined them to certain goals and which has impressed knowledge and mind on it, which is the clearest mark of divinity.[33]

Melanchthon then goes through the inner parts 'following the order of Galen', from the lower venter through to the highest. All this was taught without performing a dissection, and it is not known whether Melanchthon used any visual representations.

Twelve years later Melanchthon issued a much revised version of this book as *Liber de Anima Recognitus.*[34] In the meantime, of course, Vesalius's *Fabrica* had been published. Melanchthon acquired a copy and read it with care. At Nuremberg on 25 January 1552 he wrote a poem about the importance of the study of the human body ('De consideratione humani corporis') and placed it at the front of his personal copy of the *Fabrica*. Here it is translated:

> Think not that atoms, rushing in a senseless, hurried flight,
> Produced without a guiding will this world of novel form;
> The mind which shaped them, wise beyond all other intellects,
> Maintains and fashions everything in logical design.
> . . .
> Accordingly it follows that the body's several parts

> Came not together aimlessly as if devised by chance:
> With purpose God assigned to each its own allotted task
> And ordered that man's body be a temple to Himself
> . . .
> Wherefore, as man reflects upon the marvels in himself,
> With reverence let him venerate his Maker and his Lord,
> And keep the temple undefiled, immune from any stain,
> Lest wrath divine in vengeance come and hurl it crashing down.[35]

In his revision of his commentary on the soul (made from June to November 1552), Melanchthon spoke about the anatomy of the human body in a way which echoed St Paul writing to the Corinthians:

> looking at this wonderful variety of work and these designs of God from without and through a thick darkness, we are struck dumb and grieve that we cannot look into nature and discern causes. But then at last when we discern the 'idea' of nature in the divine mind we shall look into that whole machine as if from the inside, and we shall understand the designs of the Maker and the causes of all the divine works. Now, through this incomplete consideration, we know that God is the Architect, and we should be inflamed with desire for that perfected wisdom.

The message is that the knowledge of anatomy is for Protestants an important means to the knowledge of God: Melanchthon is saying that through anatomy we see God 'through a glass darkly' – the best that can be obtained here on earth – 'but then face to face; now I know in part, but then shall I know even as also I am known' (Corinthians I, 13,12). Moreover, Melanchthon added a long elegy on the utility of the doctrine of the fabrication of the human body ('Elegia de utilitate doctrinae de fabricatione humani corporis') by Johannes Stigelius. Melanchthon wrote in his revision that he had 'followed the best authors, *Galen*, *Vesalius* and *Leonhard Fuchs*'. Aware of the shortcomings of his work, however, Melanchthon urged anyone else who was to teach this anatomy to others to amend any errors he may have made, 'especially since there exist the figures of Vesalius, and the most excellent descriptions, about which Paulus Eberus has written thus:

> Quantum nocte alias stellas, cum lumine fratris
> Orbem complevit, vincere Luna solet:
> *Vesalii* tantum reliquis liber anteit unus,
> Corpora qui qua sint condita ab arte, docent.[36]

> As much as the Moon regularly outdoes the other stars at night, when it fills the orb of its brother [i.e. the earth] with light, so does the one book of Vesalius surpass the others which teach by what art our bodies have been fashioned'.

God is even more present in this account than in Melanchthon's earlier one, with the heart now portrayed as the domicile of God, the brain as

His temple. This work went through many editions (1558, 1560, 1562, 1569, 1571, 1574, 1584), mostly at Wittenberg. Anatomical sheets, based on the figures of Vesalius, were printed to be used with it (see Figures 8.5 and 8.6). Such teaching of anatomy in the philosophy course became common in other reformed universities too, as at Glasgow in the 1570s and Edinburgh in the 1580s, and such teaching continued throughout the

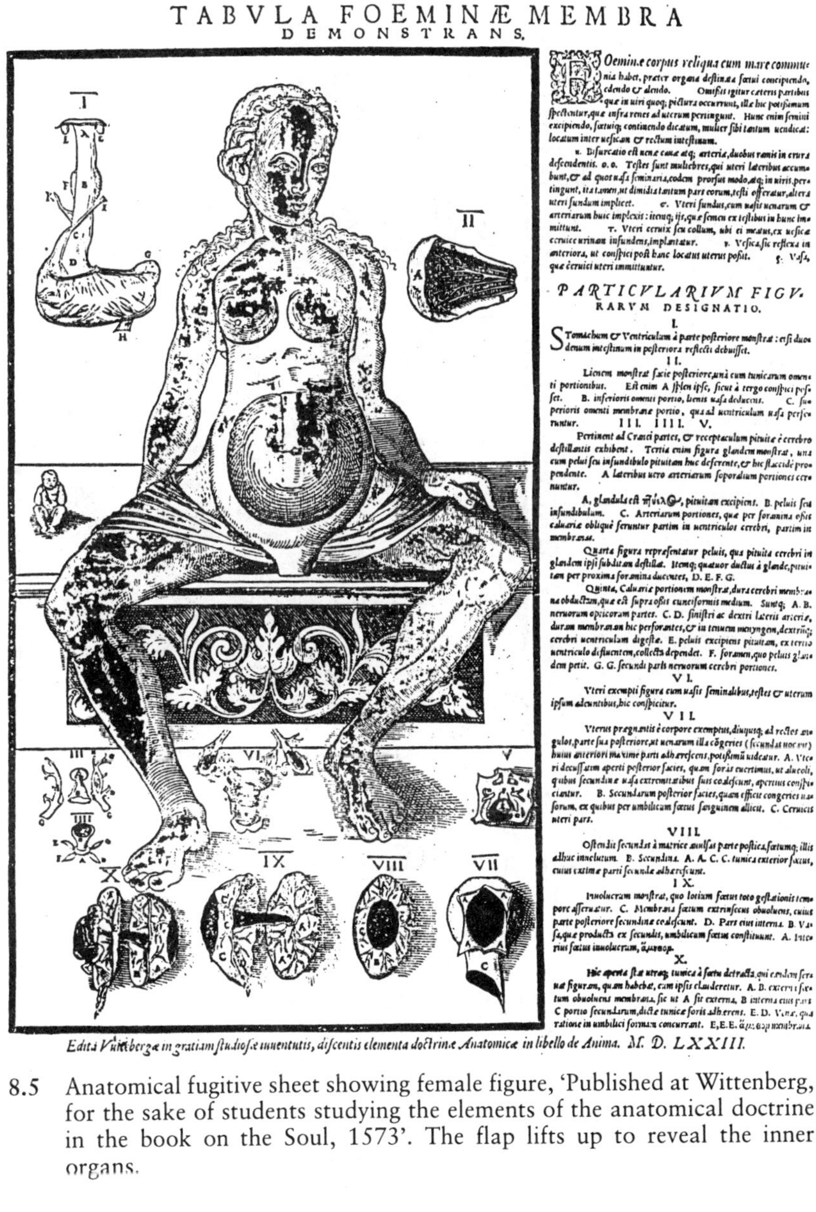

8.5 Anatomical fugitive sheet showing female figure, 'Published at Wittenberg, for the sake of students studying the elements of the anatomical doctrine in the book on the Soul, 1573'. The flap lifts up to reveal the inner organs.

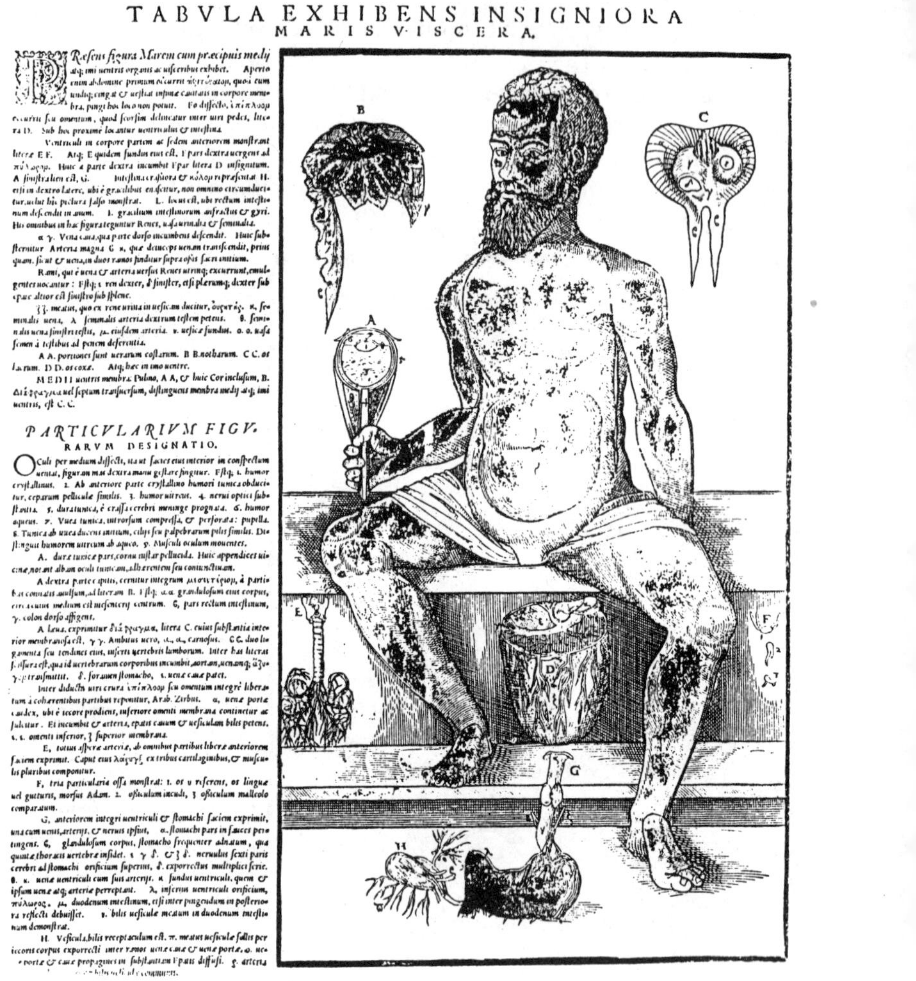

8.6 Anatomical fugitive sheet showing male figure; published at Wittenberg, probably in 1573. The figure has the facial features of Vesalius.

seventeenth century.[37] Quite independently of any concern about the advancement or promotion of medicine, Protestantism thus had a new, more philosophical and theological role for anatomical knowledge; that this was simultaneous with the endeavours of Vesalius would seem to be more than a coincidence. For the Lutheran Protestant, the human body was the temple of God, and seeing was believing.

Thus although the evidence is not conclusive, yet because, it shows, in my view, Vesalius's work in anatomical reform replicating precisely

Luther's work in religious reform, and because the Protestant leaders themselves took up Vesalian anatomy and made it central to their philosophical teaching in the Protestant cause, it points to Vesalius being a Lutheran and as coming to a new view of the body *because* he was playing out his Lutheranism into his practice of anatomy. At the very least we have Vesalius behaving here *as if* he is a Lutheran, and turning his attitudes toward reforming anatomy. And, just as Luther's work in theology can be seen as being primarily a product of the changed economic, social and political circumstances of northern Europe, so equally Vesalius's work in anatomy, though performed in the south, can be seen as a product of these same altered social and political circumstances. Indeed neither of these intellectual revolutions were events isolated from the concerns of the real world.

Of course, it might all just be a mighty set of coincidences, all the elements that the revolutionary approaches that Luther and Vesalius had in common – the pointing, the touching, the making of the senses pre-eminent as a touchstone of faith, the promotion of new texts as authoritative, the rejecting of intellectual fathers and traditions, the making of personal experience primary, the making responsibility for one's own views a moral matter. It might just be a case of two rebellious and outspoken young men who each wished to leave their mark on history in their different spheres, and who just happened to live at much the same time. It might. But in my view it is nevertheless highly suggestive of a stronger connection of a causal nature.

Luther tried to reinstate the pure practice of the pristine church, the congregation of all believers; Vesalius tried to reinstate the pure practice of pristine anatomy, the participatory engagement in anatomizing of humans as the Alexandrians had done. Luther redefined the nature of the rituals of proper worship; Vesalius redefined the nature of the rituals of proper anatomizing. Both of them believed that a proper representation of God's wonderful works in pictures was positively desirable in order to help the memory – in altar-pieces or anatomical tables. In addition, both had themselves portrayed too – and *as* apostles, as Ancients, in the very act of reviving and living out ancient pristine practices. Both of them gave a new priority to the sacred book – the Bible in the case of Luther, the human body in that of Vesalius – which, they claimed, should be preferred over any human commentator, and which was open to be read by everyone. Neither was prepared to take as doctrine anything which could not be pointed out with the finger in the sacred book, the Word, the body. And, for what it is worth, both had bitter public rows with teacher-figures of the previous generation, who thought they had gone too far: Luther with Erasmus, Vesalius with Sylvius.[38] Both Luther and Vesalius changed the nature of the debate. Both Vesalius and Luther sought to restore to modern life an ancient practice, and ancient beliefs: one of them tried to restore primitive Christianity, the other primitive anatomy. Both of them did these things in the same world.

(iii) The 'Radical Reformation'

Luther's activities constituted only the first major event in what was soon to be called the Protestant Reformation. Luther appeared wildly radical at first; but he was shortly made to seem quite tame in his approach to religion by (among others) one of his old colleagues at Wittenberg. Vesalius too appeared wildly radical in anatomy; but he too was outdone in radicalism in anatomy by (among others) one of his fellow-students at Paris. There was a new radicalism in medicine too, which had important implications for the role and significance of anatomy. There are two events of significance here for anatomy, neither of which were dealt with in the first part of this book, for although the advocates of each both looked back to the distant past, one concluded that there was *no* role for traditional anatomy in understanding true medicine, while the other, though a skilled anatomist, made his radical restatement of what goes on in human respiration solely in order to satisfy his exclusively theological concerns, and did not do this within the context of university teaching, whereby a new anatomizing tradition might be started. The new radicalism in medicine and anatomy was brought about directly by a new radicalism in religion, so in this case the connection between religious commitment and attitude to anatomy is very straightforward. We need therefore to look at some further events of the Reformation in order to appreciate the way in which they made new ways of thinking about medicine and about anatomy immediately possible, desirable and unavoidable.

Orthodoxy and heresy, whether in religion or any other area, are very volatile states. Someone's orthodoxy or heresy is not a property of what they think or preach, but of where and when they think or preach it. Positions can be reversed overnight: yesterday's heretic can become today's orthodox. It all depends on who wins the fight for power, on whose opinion prevails and thus gets transformed from opinion (where assent is optional) to doctrine or fact (where assent is obligatory). In the religious realm Luther was at first of course judged a heretic by the powers-that-were. By various chances of fate, however, Luther's cause was to prevail in certain areas. So then there would be two major Christian orthodoxies in Europe: a Protestant (Lutheran) one, and a Catholic one. But Protestantism too had its heretics. They started up almost immediately (even before the term 'Protestant' had been coined), and they were contesting with the Lutherans for the whole territory, for the confessional assent and loyalty of all Christendom. Everyone was claiming to have the whole truth. There had to be a fight; it was a fight, often quite literally, to the death. Like Luther's own claims, these new claims which Luther in his turn now judged to be intolerable deviations from the will and Word of God, had direct implications for the kingdom of this world and who should inherit it, just as much as they had for the kingdom of heaven. And like Luther himself, the advocates of these other heresies also looked backwards for their authentication; everyone was claiming to be looking back to a more primitive, a more pure, form of Christian belief and Christian life, and restoring it to modern times, and that they were engaged in God's work in doing so.

In recent decades historians have increasingly come to use the term the 'Radical Reformation' for a series of movements which suddenly sprang up and challenged the Lutheran reform from a more radical perspective.[39] We shall be concerned here with two of the groups they treat under that title: the Spiritualists and the Antitrinitarians. It will be necessary also to look briefly at another such group, the Anabaptists.

Wittenberg itself was the centre of early radical diversions from Luther's views, and one of its early leaders was Andreas Bodenstein von Karlstadt, a fellow professor with Luther, and one who disputed on Luther's side at Leipzig in 1519 against Eck. In 1520 Karlstadt wrote *Verba Dei*, and *De Canonicis*. His most recent biographer says that the first of these arose from an incident at Leipzig where Eck had asked Karlstadt whether one should say the same thing to illiterate folk that one says in sermons or theological debates. Karlstadt thought about it, concluded, and now proclaimed that, yes, everything in Scripture ought indeed to be available to all, not just to the learned! And in the second work, the *De Canonicis*, Karlstadt claimed of the Bible that 'under the letters Christ lives, breathes, speaks and explains to all in common. Therefore the interpretation of scripture will belong to all Christians . . . I want that understood thus – all to whom the Lord bestows the gift of interpreting scripture are able to interpret it, whether they be laymen or clerics, secular or religious . . .'[40] As the historian Sider puts it: 'In *Verba Dei* he urged all laymen to read the Bible; in *De canonicis*, he affirmed their right to interpret it.'[41] The sympathies of Karlstadt, the university teacher, were tending toward the common man. But he was not yet out of step with Luther.

When Luther was spirited away to the Wartburg castle by his ruler, Frederick of Saxony, after the Diet of Worms in 1521, Karlstadt was prominent in the so-called Wittenberg Movement of late 1521 and early 1522, in which the local monks (Luther's own colleagues) and the students (attracted now in great numbers to the little university) and others, excitedly tried to institute the changes they believed Luther had been calling for. They tried to change the mass (indeed to abolish it), they ran riot in the churches, throwing out and destroying the images they found there. But when Luther returned home he saw these changes as a 'theatre of Satan' (*scena Satanae*) and denounced them. Karlstadt, however, made the opposite choice: he now renounced his doctoral title and began to call himself a peasant. He left Wittenberg and went to be a local pastor, farming his own patch of land. He had joined the people. Luther pursued Karlstadt as a heretic.

A second early follower of Luther was Thomas Müntzer, a priest in Zwickau. He preached the new doctrines to the industrial artisans in southern Saxony, who were experiencing the distress of severe inflation and periodic unemployment.[42] Perhaps as a result of meeting a certain weaver, Niklas Storch, Müntzer began to teach a doctrine which was mystical, spiritual and apocalyptic;[43] calling, in the name of the imminent Second Coming and millenium, for bloody revolution and the violent overthrow of established authority. And all this was in favour of the interests of the common man, and against the Lutheran reforms which, to Müntzer, smacked too much of accommodation to the needs of the princes. Lutherans and Catholics were to collaborate in putting the radicals to the sword in order to maintain their own peace.

Most important for our understanding of their dispute (and of its relation to anatomy and medicine) is an appreciation of the different sources of authority to which Luther and Müntzer were respectively appealing. We have seen with Luther that he appeals directly to the written word of Scripture, and this is the only authority he will recognize: the Word, the written word he can point at with his finger. Müntzer, by contrast, saw this as a trick favouring the Masters:

> Our learned ones would like to bring the witness of the Spirit of Jesus into the university . . . they want to sit alone with their stolen Scripture . . . Listen, common man. You must teach yourself if you are to avoid further deception! . . . Every man should look closely at himself and see in his own life how he himself is a holy temple . . . Every man should further understand that the Holy Spirit is present as his schoolmaster in faith and be attentive to the Spirit's workings . . . The learned theologians . . . preach unashamedly that the poor man should submit to the skinning and fleecing which the tyrants have prepared for him. [They say] theologians should read the beautiful books and peasants should listen to them, for faith comes through hearing. Oh yes, there the theologians have found a fine trick . . . The people have always assumed and think still today that the clergy are authorities in the faith because they have read so many great and beautiful books. Hence the poor common man marvels: Such impressive men with their red and brown berets! . . .

> They [the theologians] poison the Holy Spirit with the Holy Scripture.[44]

Or to put it in the words of a biblical passage much quoted by men of Müntzer's persuasion, 'The letter killeth, but the Spirit giveth life' (II, Corinthians 3,6). 'Even if you gobbled up whole Bibles it would do you no good . . . you will not have faith unless God gives it to you himself and teaches it to you himself,'[45] as Müntzer proclaimed.

The source of truth lies inside God's chosen ones – the Elect – placed there by God in the form of the Spirit. Only people who have the direct, inner, experience of faith have real faith, certain faith, real knowledge of God's will. They have a direct line to God's truth and God's wishes. As Ozment has recently put it, it is not the preacher speaking to the ears (as Luther held) but the Spirit of God working within the heart that makes true believers.[46]

Mysticism-spiritualism can serve many ends. As such it was not new in this period. But the ends it was serving were completely topical. On this occasion, a spiritual turning inward (to the unworldly, the other-worldly) could be a way of turning outward – of achieving things in the outer, material world of power and human conflict. Mystics, spiritualists, are always calling on a source of authority beyond the customary: they are putting themselves beyond the judgement and values of those powerful in civil society. In Müntzer's case appeal to the workings of the inner Spirit of God enabled him to deny and circumvent the source of authority (written authority) that Luther, the university professor, held as supreme. Müntzer's 'text' was inside him, and only he had privileged access to it. This is what now made the spiritualism of Müntzer and others powerful and dangerous: the 'inspired' ones were repudiating external authority, asserting that they themselves were arbiters of their own destiny and that of society. Their inner message came from God, it was the Holy Spirit working. Their inner message told them that it was time for the Elect to rise up and eliminate the ungodly, to cleanse the world ready for the Second Coming. It was time for the common man to kill the princes and their agents, and to enter on his divinely ordained inheritance.

This spiritualism had a wide appeal to the disaffected artisans and peasants among whom Müntzer and Karlstadt threw their lot. It is part of the inspiration, and also a source of some of the language, of the Peasants' War of 1524–25, which was bloodily put down by the Masters.

The teachings of Thomas Müntzer may have helped inspire another major radical religious movement, that of the Anabaptists, a rather ragged and disorganized set of small sects. They are supposed to have sprung up first in January-June 1525 (in Zurich), and they too were seen as devils and enemies of civilization by both Lutherans and Romanists. Their strategy of radicalism was different from that of Müntzer and the peasants. For they sought to separate themselves from the world; to create a community of believers and saints, consisting entirely of voluntary members, living according to their understanding of Christ's teachings – 'a congregation in which repentence was in evidence by newness of life in Christ', as an early Anabaptist described it.[47] They had a commitment to holy living – to live as they believed Christ and the disciples had lived – based on brotherly love, discipleship and pacifism. All this, in their eyes, entailed separation from the rest of society. This separation was taken by others to be destructive of Christendom, and of all peace and morality. The distinctive mark of an Anabaptist was that he or she had been re-baptized (*ana-* = anew, again) as a consenting adult. For Anabaptists infant baptism was of no effect; only adult baptism consciously and deliberately undertaken could be of significance. Yet infant baptism (and on this both Romanists and Lutherans agreed) had of course hitherto been the most fundamental ritual by which one became indissolubly joined to the social and political family of Christendom; it was one of the traditional sacraments. To renounce infant baptism was to renounce both traditional Christianity and traditional Christendom. To withdraw from traditional Christendom was to set up a rival – and hence threatening – form of society. To seek to build a society which practised the holding of property in common was to challenge the very basis of traditional society. So dangerous was the Anabaptist poison taken to be, that the Anabaptists were hounded and horribly put down by both Protestant and Catholic princes and by city magistrates of all confessional hues, and their New Jerusalem, their experiment at Münster in communist living (1533–35) – based on their understanding of the way of life of the primitive church – was attacked until not a single person survived.[48]

Meanwhile the old imperial and free cities, in Germany and especially in the Swiss cantons, under pressure from lords without and from guilds within,[49] were subject to intense unrest stemming from the reformation movement. It is in some of these that the dramas of the radical reformed medicine and anatomy that we are here concerned with were to be played out. In general the northern towns became Lutheran and the southern ones adopted the approach and creed that Ulrich Zwingli first fashioned for Zurich in the early 1520s.[50]

In all the towns reformation was initiated 'from below'. The stages of reform over the years *c.* 1520–*c.* 1535 followed a general pattern, consisting of calls for reformed preaching, followed by riots, iconoclasm, petitions – all from below. These were met – from above – by staged public disputations, the replacement of the mass by evangelical communion, the closing of the local monasteries, the adoption of reformed principles in worship, and the creation of a new form of urban confession. In cities where there was a university, the professors were usually found to be conservative and against reform; they could be replaced by more congenial teachers.

In all the towns the magistrates took very seriously their newly acquired religious functions. The town – all its members – was to be of one faith and confession, and of one discipline. Calvin's Geneva (1536–38 and 1541–) with its *Ordonnances Ecclesiastiques*, drawn up by Calvin in 1541, is only the best known and most severe of such new rules for living in an urban Christian community. Each such reformed town was a little self-contained church, a theocracy in the sense that the magistrate and pastor together worked to realize the rule of God there.[51] And this, in every case, was supposedly a restoration of Christian living as in the primitive church.

With the 'Radical Reformation' came a radical medicine. It was in the towns of the Swiss cantons, precisely while they were undergoing the throes of reformation, that there appeared someone preaching a radically new kind of medicine, and one which therefore made a new kind of body visible. It was Philippus Aureolus Theophrastus, better known by the fantastic name he gave himself: 'Paracelsus', 'greater than Celsus'.[52]

Paracelsus was born in 1494 in a small town of the Swiss cantons, Einsiedeln, south of Zurich. The son of a physician, Paracelsus was brought up among mining folk in Einsiedeln and then Villach (further east in the Alps), in the same sort of depressed mining areas where Thomas Müntzer found his constituency of poor distressed artisans and peasants. Paracelsus was later (in the 1530s) to write on miners' diseases.[53] Like Müntzer and Karlstadt, Paracelsus was a populist Spiritualist.[54] Like Müntzer and Karlstadt, Paracelsus came to this position in the 1520s, on his return from a number of years wandering abroad. In his wanderings he may have acquired an MD from Ferrara University; like Karlstadt, the university professor, after his conversion Paracelsus too was to reject and denounce the learning of the Schools. Like both Karlstadt and Müntzer, Paracelsus was active in the Peasants' War of 1524–25 on the side of the peasants.

Like Karlstadt and Müntzer, therefore, Paracelsus was hostile to the Word as wielded by the leaders of the 'Magisterial Reformation' such as Luther. He was hostile to book-learning, to Latin, to logic. Instead, and this is yet again just like Karlstadt and Müntzer, Paracelsus put his trust in the Spirit, in the Word within, in holy inspiration. Though not a priest, he preached extensively, inspired by the Spirit. Like Karlstadt and Müntzer, Paracelsus believed that only people who have the direct, inner, experience of the Spirit have real knowledge of God; and only those who have direct, inner, experience of things (*res*) have real knowledge of things. And this meant that when he considered medicine, Paracelsus was to put a new stress on the importance of personal experience of things, personal experience of a particular kind: *spiritual* personal experience. For only the Spirit gave true knowledge, in this domain as much as in the other.

It was in the spring of 1527 that Paracelsus arrived in Basle, a city in the midst of its reformation. Allying themselves with the 'democratic' elements represented by the craft guilds, were the religious reformers in Basle, led by Oecolampadius, who had been appointed professor of theology in the town university by the town council. The Anabaptists appeared in town in 1525, and staged disputations against them were arranged.[55] There was a revolt of the town's weavers in the same year. The town council secularized the monasteries and allowed the mass to be substituted by evangelical communion. As pressure for reform grew the university was found to be on the conservative side, and the town council tried to mediate between it and the reformers.[56] It was at this point that Paracelsus was invited 'by an ample stipend' from the town council to become a professor of medicine in their university, as part of their scheme to reform it.[57] Paracelsus's arrival thus coincided with the height of democratic demands, and his short stay (less than a year) was to be followed by iconoclastic riots, calls for annual elections to the town council, and the enforced closure of the university.

During his stay (and also after he left Basle) Paracelsus preached a medicine of the people, an anti-elite form of medicine. He preached it in the language of his people: Swiss German. On 5 June 1527, he gave a public speech which, it is said, he accompanied by publicly burning a volume of Avicenna, the great Arab Galenist physician. This speech (or rather, the advertisement for it) is the famous 'Basle Programme', and it announced Paracelsus's Spiritualist reform of medicine:

> Since of all the disciplines medicine alone, as a kind of divine gift, is praised by both sacred and prophane authors by being called 'necessary'; and since very few of the learned teach it properly today, it seemed good to recall it to the pristine reputation of its own authoritativeness and, in so far as something may be retrieved from the dross of the barbarians, we shall purge medicine from the worst errors, giving our preference not to the precepts of the Ancients, but only to those precepts which we have discovered – partly from the signs of its own nature (*indicatione rei naturae*), partly by our own efforts – and have proved by the long use of things and by experience (*longo rerum usu atque experientia*). For who does not know that most of the learned in this age have lapsed most disgracefully, to the greatest danger of the sick, since they have adhered far too anxiously to the sayings of Hippocrates, Galen, Avicenna and others just as if they originated from the

Delphic seat, like oracles, from which no-one were allowed to depart a finger's breadth. From these authors the most splendid scholars may indeed be produced (God willing), but not physicians. It is not a title, nor eloquence, nor skill in languages, nor the reading of many books (even if these add not a little ornament) which is desirable in a physician, but the highest knowledge of things and mysteries (*summa rerum ac mysteriorum cognitio*), which one kind of knowledge stands easily in place of all the others. It is indeed the role of the rhetorician to be able to speak clearly and to persuade and sway a judge to his opinion, but it is the role of the physician to have learnt the kinds, causes and symptoms of diseases and moreover to apply drugs to them by his sagacity and industry and, according to the spirit and nature (*ratio*) of each remedy, to heal all diseases.

In order to explain in a few words my mode of teaching, it is necessary to speak of myself: invited by an ample stipend from the councillors of Basel, I expound for two hours a day the books, both of active and theoretical medicine and of physics and surgery, of which I am myself the author, with great diligence and to the great profit of my listeners; but I have not scraped these books together from Hippocrates, Galen or other authors, as other people do, but I have acquired them from the greatest teacher of things (*rerum*): experience and labour. Similarly, when I want to prove anything, experiment and *ratio*[58] supply for me the place of authorities. For which reason, dear readers, if the mysteries of this Apollonian art delight anyone, and love and desire for it sieze him, and he desires in just a short time to learn whatever there is of this discipline, he should betake himself forthwith to us at Basel, and he will learn things much other and greater than I could describe in a few words. But so that our plan should be known more clearly to students, it is no trouble to append here as an instance, that we never imitate the ancients in the matter of complexions and humours, whereby all diseases bear names mistakenly based on them; whence it is the case that today diseases, causes and critical days have been accurately learnt by none or very few scholars. I permit you to judge nothing hurridly of these claims till you have first heard Theophrastus. Farewell, and reflect good men on this our attempt at restoring medicine.[59]

The new medicine is not something that can be learnt, but is a gift; it is available to all who have the gift, not just to the elite. It will be based not on books and reason, which may produce rhetoricians but are useless for producing physicians. It will be based not on the authority of the Ancients – the Word without – but on one's own experience, the Word within.[60] It will be based not on logic and learning, but on mysteries and things (*res*).

The 'experience' of which Paracelsus here speaks is not the same kind of experience that Vesalius was to call for. For to Paracelsus 'experience' is mystical experience, and the 'natures of things' that he seeks are mystical natures. They are available only to those in whom the Spirit is active. At the Fall, when Adam and Eve were expelled from Eden, God gave man 'the light of nature' – the true reason (*ratio*) so that he could understand created nature.[61] Paracelsus says,

> it is nature that teaches all things, and what she herself cannot teach, she receives of the Holy Ghost, who instructs her. For the Holy Ghost and nature are one, that is to say: each day nature shines as a light from the Holy Ghost and learns from him, and thus this light reaches man, as in a dream. Everything that comes from the light of nature must be learned from the light of nature.[62]

All this is Spiritualist: medicine and medical knowledge for Paracelsus are the product of inner experience. It is necessary to look at and inspect the things themselves of the world, the diseases and the possible remedies. But the 'experience' which tells you about their natures, is an inner experience, literally an insight. And, according to Paracelsus's teaching, the thing in Nature that one seeks as a remedy is an 'arcanum', literally a 'secret' or 'mystery', or 'a little heaven'. It is the physician's role to extract this active part – the spiritual part – of things from their material dross. Thus, what the Paracelsian physician deploys are spiritual forces as cures for diseases – diseases which can themselves only properly be understood as spiritual entities, as the operation of spiritual forces.[63] For Paracelsus (as Pagel has well put it) 'the substances which we handle in daily life are but crude covers that envelop and disguise a pattern of spiritual forces'.[64] Recognition of these forces in plants and minerals is via signs, symbols, signatures – all visible to the inspired physician. The most important means for releasing the spiritual power of the active ingredient from the matter in which it is hidden is, for Paracelsus, alchemy.[65] And in making separations of the spiritual from the material by alchemy, the physician is assisting in God's work, the divine work of helping perfect and purify things for their final, spiritual end.[66]

This is Spiritualist in many ways: in the kind of things Paracelsus claimed diseases and remedies to be; in the nature of the physician's activity; in the source of his proper knowledge. But it is Spiritualist in an even more fundamental way, in that it is explicitly and deliberately Christian. It was the first 'Christian' form of medicine ever established (though Paracelsus, of course, took it for granted that he was restoring an ancient practice). And this is one of the reasons that Paracelsus was so opposed to the Greek rationalist tradition in medicine, logic, philosophy or rhetoric: the Greeks were not Christian. Proper Christians, as Paracelsus's example shows, ought to be seeking kinds of knowledge which deal in the spiritual, the mystical and the mysterious, not in the rational and the world of cause-and-effect. Only false Christians (the

Masters) use the outer Word of the Greek authorities. Paracelsus saw himself as restoring a truly Godly and Christian medicine, the only kind of medicine that a proper Christian should follow. It is a medicine by, of and for Spiritualist Christians. God has given his proper followers the inspiration to read the book of Nature and thus recognise the cures they need. The proper – Christian – physician looks inward for his learning, which is placed within him by God. In constructing his medical approach Paracelsus was, as recent historians have shown, deeply indebted to various ancient Jewish and neo-Platonic traditions, with respect especially to astrology, magic, the astral body, Hermeticism, and microcosm-macrocosm analogies, all of which are evident throughout his writings.[67] But Paracelsus's medical approach was more than just a synthesis of such traditions and borrowings. And its seemingly 'proto-scientific' characteristics, which have also concerned historians, such as its empiricism and stress on personal experience of things, its interest in experiment, its development of chemical techniques, and its anti-authority stance, were consequences rather than the source and cause of it. For Paracelsus's medical approach was constructed in the cause of building a Spiritualist, Christian medicine which would replace Greek rational medicine. It was a radical medicine deliberately opposed to 'Magisterial' medicine, deliberately based on invisible inspiration rather than on the visible Word. For the letter killeth, but the Spirit giveth life.[68]

For our present purposes it is not necessary to expound Paracelsus's medical approach in any greater detail. For it should be obvious by now that in Paracelsian medicine there was no room whatsoever for the anatomical tradition deriving from Galen, and which Guenther and Sylvius (and also Vesalius) devoted themselves to recover, nurture, preach and practise. In this Greek tradition, anatomy was treated as the fundamental underpinning, both literally and metaphorically, of Galenic medicine. With the rejection of the heathen medicine of Galen, Paracelsus also rejected its anatomical tradition:

> It astonishes me that you set up the dead body as a ground of knowledge of what is useful for the living body, without considering that the essence, property, being and spiritual force (*Kraft*), which is the highest part of anatomy, have died and become corrupted . . . Thus I warn you that if you ever wish to preserve the living body, not to seek its benefit in the dead.[69]

Paracelsus's views on harmony and wholeness did not accord with the Galenic views of part-ness and cutting. The relation between cause and effect was quite different in the two approaches; where the Galenist looked for a geography of physical connections (structure, action, use), Paracelsus looked for a network of sympathetic relations – connections spiritual not physical. A Galenist like Vesalius saw with his bodily eyes, Paracelsus with his spiritual eyes – by the light of Nature. For Paracelsus, intuition and faith held the place that pointing with the finger and witnessing with the eyes held for Vesalius. The human body that Paracelsus saw bore no relation at all to the body that Vesalius saw. The body that Paracelsus saw could not be captured in a physical image.

The kind of anatomy that Paracelsus favoured was quite different. A contemporary, who was by then a Lutheran, later wrote that Paracelsus claimed that there existed

> a brief and simple anatomy of the human body, more useful and commodious than that prolix one of Galen; especially if one enquired to what extent the little world, that is man (whom Aristotle in his language called *microcosmos*), paralleled the larger world; for he proclaimed that this was the most firm basis of the whole of medicine.[70]

Pagel and Rattansi have explored Paracelsus's various uses of the term 'anatomy', and demonstrated how 'Paracelsian anatomy is the study of the parallelism between the greater world and the lesser world, the macrocosm and the microcosm.'[71] This is nothing other than a Spiritualist anatomy.

Obviously, everything about Paracelsus's approach both to medicine and to anatomy was antithetical to the learned Galenic physician of Paris or Padua. Paracelsus treated their authorities as dross, and he repudiated the efficacy or desirability of the whole tradition of their rationalist medicine. He could not simply be ignored, for by all accounts his medicine worked: it cured people. Many converts, both physicians and patients, were made to the practice of Paracelsian medicine, especially north of the Alps. It was to remain a powerful alternative medicine for a century and more, appealing to the poor, the artisan, the person hostile to established authority and monopoly; but most of all the medical doctrines of Paracelsus appealed, and continued to appeal, to radical protestants – to followers of religious approaches similar to that of Paracelsus himself.[72] Although one or two people could, by the 1570s, make some accommodation between Galenic and Paracelsian medicine,[73] it was to take a long time to draw the sting from the radical threat to Galenic medicine and traditional society that Paracelsus preached.

It was the 'democratic' and populist nature of the Radical Reformation which now led Paracelsus the physician to conceive and preach a radical new medicine, ensured that it was mystical and spiritual and enrolled eager ears to listen to it. Within Paracelsus's new medicine, the role of anatomy in the traditional sense was nil. But the Radical Reformation made at least one other new body visible, and to this we now turn to see the very direct relation between personal religious commitment and anatomical vision.

This other new body was made visible by Michael Servetus (alias Michel de Villeneuve or Villanovanus, from the name of his hometown; alias Reves, which was also the name of his family house there) who lived from 1511 to 1553.[74] To the modern eye, the most striking feature of this new body brought to visibility by the Radical Reformation was that it had the so-called lesser circulation in the human, that is to say, it had the blood being passed from the right ventricle of the heart to the lung, and then to the left auricle and then ventricle. 'Lesser circulation' or 'pulmonary circulation' are in fact misnomers, for the route described is not a circuit; 'pulmonary transit', or passage through the lung, is a more appropriate term. Historians of anatomy have been interested in the 'discovery' of the pulmonary transit as they have seen it as an important step or stage on the glorious road to the discovery of the circulation of the blood by William Harvey in the early seventeenth century. It has thus been a matter of concern to such historians to identify who made this heroic first step, and their primary concern with Servetus has been with allocating the right degree of credit to him.[75] Such posthumous awarding of merit marks is not our concern here, and it is particularly fatuous in the case of Servetus. For although Servetus was very famous in his own lifetime and thereafter, his account of the pulmonary transit was not known at the time to other anatomists, since it was announced in a book which was read by only a handful of people (none of them anatomists) before it was declared heretical and was burned at the stake together with its author. Only three copies are known to have survived out of the thousand printed, and it was not until the 1690s (long after the circulation of the blood had been discovered by Harvey) that anyone noticed the account of the pulmonary transit in any of these three.

If, like Servetus, you see the body as having a pulmonary transit of the blood, then you are seeing a different body to the one portrayed by Galen and which, as we have seen, Guenther, Sylvius and Vesalius were engaged in bringing back to view. For it is a body which, for the first time, has an open connection between the two blood systems of the Galenic body: the nutritive blood system (based on the liver and right heart) and the respiratory blood system (based on the lung and left heart). The implication of this, though it is not an implication that Servetus pursued, is that there is only one system of the blood, not two, and that therefore the blood must circulate around the body. But it was not a matter of Servetus 'failing' to see this implication. For although he was highly skilled at anatomizing, his opinion that the blood crosses through the lung from right heart to left did not come directly from anatomical investigation, but from theological investigation. It was, for Servetus, an essential tenet for the theological case he was arguing: to make the theological point, he called on his anatomical knowledge. The theological point he was making was very radical indeed, so radical that it has been justly described as 'total heresy'.[76] It was actually Calvin who burnt Servetus to death in Geneva, but it could just as well have been any other reformer, Lutheran or Swiss – or indeed the Catholic Church, whose Inquisition had already condemned Servetus to death but, as Servetus had escaped from custody, had had to content itself by burning him only in effigy. Servetus's heresy was so total that it united the warring Roman and reformed churches in hostility against him – which meant it was perceived as a threat of the same proportions as that presented by the Anabaptists.

Had his intended readers been able to read Servetus's last book, and had they reached the passage where the pulmonary transit is described, what would have struck them most would probably have been not the transit itself and its implications for Galenic theory, but the context in which it was described; a context which would have looked pantheistic, promoting the belief that God is personally present in every part of His universe, in every stone, in every tree, in every part of the human body, and as implying psychopannychism – the temporary death of the soul until the resurrection. This would have been rejected by almost everyone as being completely alien to the orthodox Christian views of the soul and to the Aristotelian and Platonic views of the distribution of soul in the body. Servetus's opinion that the blood passed through the lung would thus not have received a hearing even if there had been ears to hear.

Servetus came from Spain. In 1492 the non-converted Jews had been expelled from Spain; the converted Jews remained behind, but were called by the contemptuous title of 'Marranos'.[77] Most difficult of the Christian doctrines for Jews to accept were those of the Trinity and the deity of Christ: to Jews it sounded as though the Christians had three gods, whereas they knew there was only one true God. Servetus travelled in the service of a Franciscan theologian (1525–29), and he took time to study law at Toulouse (*c.* 1528–29). He was then taken by his patron to Bologna where he witnessed the papal coronation of the Holy Roman Emperor there in 1530. At some early period he mastered Greek and Hebrew. While still in his teens, when Lutheran biblical literalism was at its height, Servetus adopted this new fashion of looking to find the biblical warrant for Catholic doctrines. On looking in his Bible he could find no mention of the Trinity, that supposed cornerstone of Christian belief – and major impediment to the conversion of the Spanish Jews to Christianity:

> Not one word is found in the whole Bible about the Trinity, nor about its Persons, nor about an Essence, nor about a unity of the Substance, nor about one Nature of the several beings, nor about their other babblings and disputes of words, which Paul says belong to the knowledge which is falsely so called.[78]

Erasmus, Luther, Melanchthon and others were aware of this absence, but had halted at pursuing it to its logical implication; instead, they chose to cling to the Trinity (as they also did to infant baptism) as a most fundamental article of faith, a choice which indicates the limits on reform that they were prepared to countenance.[79] But for Servetus, this astonishing absence from the Bible of such a central article of traditional Christian belief meant that it was an obviously false belief. He turned to investigating how it had been imposed on, and had thence corrupted, Christianity. From his reading of the early Church Fathers he eventually found the Council of Nicaea to be the guilty party; from their promulgation of this doctrine (in 325), Servetus dated the fall of the Church and the growth of the temporal power of the papacy. What Servetus henceforth wanted to do was to convince his contemporaries of the falseness of belief in the Trinity, in order to give them a correct understanding of the nature of God – and hence promote the restoration of the Church to the condition it had been in before the Fall produced by the Council of Nicaea. At a mere 20 years of age he burst into print with *On the Errors of the Trinity*, of 1531, 'by Michael Serveto, alias Reeves, a Spaniard of Aragon'.

The advocates of this position (of whom Servetus was the first in the sixteenth century) claimed that the denial of the Trinity 'was simply a revival of primitive Christianity, a return to the original pure teaching of Jesus and his Apostles and of the Ante-Nicene Fathers, which had for twelve centuries been obscured and corrupted by admixture with pagan doctrines of Greek philosophy'; but its opponents 'contended that the new interpretations of Christianity were only a modern revival of ancient heresies long since condemned and abandoned'.[80] This was the response that Servetus's first book met: it was attacked as a revival of the third- and fourth-century heresies of Arius and others, that is, of those whose opinions were rejected and anathematized at Nicaea. Servetus had already failed to convince Oecolampadius, reformer of Basle, of the rightness of his views in person; now Bucer, reformer of Strassburg, attacked his published views loudly. Luther and Melanchthon read and disapproved of Servetus's book: 'In Servetus there are many marks of a fanatical spirit,' as Melanchthon wrote.[81] And acting for the Catholic Church, the Inquisition started to chase Servetus too. Servetus's response was to publish a revised version of his work, the *Two Dialogues on the Trinity*, 1532. Thereafter, for a number of years, Servetus felt it was wise to go under an alias as Michael de Villeneuve (Villanovanus).

Servetus's life was now spent mostly at Lyons, but with periods of residence at Paris (1532–34 and 1537–38)[82] as he had decided to become a physician, possibly as a result of having made the acquaintance at Lyons of the celebrated local physician, Symphorien Champier.[83] At Paris, probably in 1534, he apparently had an (unkept) appointment with the young Jean Calvin, who was then involved in the notorious Affair of the Placards. He also wrote a work *On Syrups* (1537), and a *Defence of Astrology* (1538), which led him into trouble with the medical faculty.

At Paris Servetus learnt anatomy. His teacher was none other than Guenther. He became Guenther's other most valued assistant, after Vesalius. Whether or not Vesalius and Servetus knew and worked alongside each other is not clear.[84] Guenther writes approvingly of Servetus in the Basle edition of his *Anatomical Institutions* of 1539:

> The first of my helpers was Andreas Vesalius, a young man marvellously diligent in anatomy, an exemplary teacher of a purer medicine ... And after him, Michael Villonovanus amicably applied himself to dissections with me; a man well-versed in every branch of study, in Galen he is second to none. With the help and assistance of these two, who at that time wanted to hear me lecturing on Galen, I examined and showed to students all the muscles, veins, arteries and nerves in those bodies, both of the limbs and of the other parts.[85]

This praise puts Servetus's anatomical accomplishments into very distinguished company.

Through these years of wisely self-imposed exile, Servetus continued to be obsessed with the issue of the Trinity. He engaged in a correspondence on the topic with Calvin at Geneva; both were writing under pseudonyms, but each was aware of the other's identity. And eventually he published, anonymously – but with his initials printed at the end, and with the letters to Calvin included, so that he could hardly avoid detection – what turned out to be his final work, *The Restitution of Christianity*, 1553. He was betrayed to the Inquisition (rumour has it he was betrayed by Calvin), which put him on trial at Vienne, near Lyons. Servetus escaped, but the trial continued and he was condemned to death for heresy and was burnt in effigy until such time as he could be burnt in person. And then, exhibiting again the death-wish that had already brought him into so much trouble, Servetus went to the worst possible place, to Geneva, where he attended church service. His fate meant that he was inevitably recognized, and Calvin had him tried before the city council, with Calvin himself as the chief prosecutor. As

Calvin had earlier announced that if Servetus were to come to Geneva, then 'if my authority is of any avail I will not suffer him to get out alive',[86] the verdict was something of a foregone conclusion, and Servetus was found guilty of preaching against the Trinity and against infant baptism (both of them recently revived as capital offences under the Justinian Code), and was duly burnt alive on 27 October 1553, with a copy of his book strapped to his arm. Servetus was doomed to have his lust for martyrdom fulfilled since his trial turned out to be a central act in the local politics of the Reformation in Geneva, and thus, in order to retain control of the reformation of the city against the 'Libertines', Calvin was obliged to suppress any inclinations he may have had to be tender towards Servetus and to push for a guilty verdict and the death penalty.[87] Shortly thereafter Calvin himself became the subject of printed attacks for intolerance in burning Servetus as a heretic;[88] from this controversy dates Servetus's probably unmerited reputation as an advocate of freedom of thought, and Calvin's as an enemy of toleration.

Servetus was a central figure in the Radical Reformation even though, unlike Paracelsus, he was not a populist, nor did he have any known followers. He was an opponent of the papacy, and of the reformers both at Wittenberg and in the Swiss cities. He claimed to be acting under inspiration, writing (*c.* 1542), 'when I was a young man, scarcely twenty, I was moved by a certain divine impulse to treat of this cause [i.e. the Trinity], having been taught by no man'.[89] He took the position that the Bible mysteries expound themselves to those who have eyes to see, claiming that the conventional view of the Trinity (which he called 'a Cerberus', the three-headed dog which guarded the entrance to hell) was the invention of the philosophers, who 'have contrived an imaginary Trinity, three beings in one Nature', taking 'great pride in Platonizing, by multiplying separate beings'.[90]

'This plague of philosophy was brought upon us by the Greeks,' wrote Servetus's for they above all other men were most given to philosophy; and we, hanging upon their lips, have also become philosophers.'[91] But 'Christ was not speaking to philosophers . . . but to the common people, to children and women'; as for Servetus himself, he would rather speak 'after the manner of common men and not go beyond Scripture, than philosophise foolishly'.[92] He believed that the millenium was imminent,[93] and wanted the overthrow of the Church as currently instituted and its replacement by a fellowship of the spirit – a 'restitution of Christianity'.[94] He was accused of being an Anabaptist, and although at his trial he said he disagreed with the political views of the Anabaptists, he certainly agreed with them in being against infant baptism and in favour of adult baptism only. Finally, his account of God, the created world, religion and everything was highly spiritualistic.

What was it like, the new body that Servetus's radical religious obsessions made visible to him? To understand this, we need to look at Servetus's writings of 1531 and 1532 on the Trinity. Servetus was not himself against a proper concept of the Trinity, and the position he was advocating was that God showed Himself in three *dispositions*, as God the Father, Christ, and the Holy Spirit. Influenced by the philosophers with their three 'beings', the theologians had insisted on the separateness of God the Father, Christ, and the Holy Spirit (three beings in one substance). Servetus, by contrast, argues passionately for the common nature of these three dispositions of the one God. Hence he devotes many pages to showing that Christ, while still being God, really and fully participated in human-ness: Christ was the human disposition of God. Similarly he argues vehemently that the Holy Spirit too is God under another guise: it is not a separate being but, as the Scriptures say, 'an activity of God, a kind of in-working or in-breathing of the power of God'.[95]

> Of the Holy Spirit I have already said that God gives us his Spirit only in this way, that he gives us the breath of life. For life is not derived from us, nor from our nature, but is given according to the grace of God; and by the breath of God upon a mass of clay man is made a living soul ... The energy and life-giving spirit of the Godhead are in that substance which is breathed in and out; for he by his spirit keeps the breath of life in us, giving breath unto the people that are upon earth, and spirit to them that walk therein ... By the moving of his Spirit are the powers of the heaven moved. The matter of which the world is made is a lifeless thing unless it were kept in motion by the Spirit of God ... I know that God's universal Spirit fills the earth, encompasses all things, and produces the power in every man. With the prophet would I cry out, O Lord, whither shall I go from they spirit? since neither above nor below is there any place without the Spirit of God ... For it is the property of the Spirit of God to quicken and strengthen; and as no thing is made without his Word, so there is no thing, nor stone, nor plant, which has any power without the Spirit of God.[96]

Like all other parts of creation, we are maintained in life only by the constant renewing activity of God breathing life into us; the air we breathe is (or contains) the Holy Spirit – which is itself a form or disposition of God Himself. The Holy Spirit, once breathed in, then breaks out into works:

> there is a proper and spontaneous cause of works, even apart from any faith and love whatsoever. There is a movement of the Spirit which breaks out freely into works, apart from any willing or inward quality; for the outward act is governed by the movement of the heart alone which sends its spirits to the members.[97]

Of course Servetus did not come to these radical and seemingly pantheistic views simply from the logic of his argument against the three beings of the Trinity. He obviously first held certain views, for instance the view that God is everywhere and that everything is constantly sustained by God, and from this was led to see what he called the 'Errors of the Trinity'. Where he acquired such a radical and heretical view, and at such an early age, is unknown, though the Spiritual Franciscans may well be held under suspicion. Servetus's view (like that of the Spirituals) involved a belief in man himself being able to become God, by participating in the deifying Holy Spirit of God. God became man in Christ:

> One sole hallucination the philosophers had, which deprived them of this knowledge [of the humanity of Christ]. They say that the Substance of God can not be mingled with the Substance of man. Oh, pitiable madness! What else is the mystery of the incarnation but a mingling of man with God?[98]

And man can also become God. As Servetus wrote in *The Restitution of Christianity*, 'God can communicate the fulness of His deity to man, give him His divinity, His authority, and His glory'; or again, 'We are truly made Gods; by participation in the deity of Christ, we are made true participants in the divine nature.' The agency through which this transmission of divinity – this deification – is accomplished, is the Holy Spirit.[99]

All these obsessions and aberrant beliefs Servetus still nurtured over the years of hiding under his pseudonym. *The Restitution of Christianity* of 1553 deals with them all once more. And it is in this context, of arguing that God as the Holy Spirit enters constantly into us to vivify and deify us, that Servetus offered an account from anatomy in order to give his reader the full details of the soul and spirit. He had, of course, in these intervening years been a medical student under Guenther and learnt Galenic anatomy. He had also, apparently under the influence of Symphorien Champier, become much more sympathetic to one kind of ancient philosophy, neo-Platonism.[100]

Hence in 1553 he was able to speak of 'the divine philosophy which you may easily understand if you have been trained in anatomy',[101] and this divine philosophy shows that there are not three spirits active in the body as has been customarily believed (the natural spirit from the liver, the vital from the heart, and the animal from the brain), but only one – in the sense that 'in all of them is the *energia* of the one spirit and light of God'.[102] The natural spirit (as well as the animal spirit, as Galen taught) is produced out of the vital spirit in the heart, for

> the soul, breathed into Adam by God, comes earlier into the heart than into the liver, and is communicated from the heart to the liver ... The matter of the soul (*animae materia*) is made from the

> blood of the liver, through a remarkable elaboration, of which you shall now hear. Hence it is said that the soul is in the blood, and the soul itself is the blood or the sanguineous spirit. For the soul is not said to be principally in the walls of the heart, or in the body of the brain or of the liver, but in the blood, as God Himself teaches in Genesis 9, Leviticus 17 and Deuteronomy 12.[103]

What happens is that the vital spirit is 'generated from a mixing, made in the lung, of the inspired air with the elaborated subtle blood which the right ventricle of the heart communicates to the left'.[104]

> This communication is not made through the central wall [the 'septum'] of the heart, as is commonly believed, but the subtle blood is driven from the right ventricle of the heart by an excellent contrivance (*magno artificio*) by a long route through the lung. It is prepared by the lung, it is turned reddish-yellow: and it is transferred from the vein-which-looks-like-an-artery into the artery-which-looks-like-a-vein. Then in this artery-which-looks-like-a-vein it is mixed with the inspired air, and purified of fulgineous vapour by expiration. And thus at length the mixed whole is drawn by diastole by the left ventricle of the heart, suitable stuff to become vital spirit.[105]

This, then, is how the Holy Spirit enters man and acts within him. Instead of a little blood seeping through the septum of the heart from the right ventricle to the left (as Galen held), lots of blood flows from the right ventricle along the vein-which-looks-like-an-artery to the lung where it is mixed with the Spirit, and this mixed blood is then drawn along the artery-which-looks-like-a-vein to the left ventricle of the heart, from whence it passes to the arteries of the body. Servetus was perfectly aware that the account he was offering did not accord with that of Galen or, as he put it, it was a truth 'unknown to Galen'.[106] But it necessarily existed for theological reasons. Hence it necessarily existed on anatomical grounds too: 'the significant size of the vein-which-looks-like-an-artery confirms this, which would not have been made thus or so large, nor would it emit from the heart itself such a force of the purest blood into the lung just for nourishing it'; similarly this very copious mixing (*tam copiosa mixtio*) must take place in the lung rather than in the left ventricle, as the ventricle is not large enough; the septum of the heart could not have little pores in it to allow the blood to seep through, and so on.[107] In other words, Servetus was now able to see in the body the structures and pathways that his devotion to the constant in-breathing of the Spirit demanded. It is likely that he could see these even when he had been dissecting with Guenther. Servetus here makes no wholesale rejection of Galenic anatomy (as Paracelsus did), but the body he sees – pervaded, vivified and infused with the Holy Spirit – is nevertheless vastly different from the Galenic one.

All that Servetus says, both about the Trinity and about the blood, is based and built on the literal text of the Bible. In particular, his belief that the blood has to pass to the lung (to be mixed and imbued with the Holy Spirit) is based on a literal reading of the biblical words that 'the soul is in the blood' (Leviticus 17, verses 11, 14). For, given this, Servetus found it necessary to be able to specify precisely where and how the soul or spirit came into the blood. Going back to the Bible, and going back to reading the Bible in a pre-Trinity light, made a new body visible to Servetus, one not visible to his contemporaries. It is a body which is full of the Holy Spirit, breathed in with every breath, the Holy Spirit which is sent to the parts of the body by the heart and thence breaks out into good works. For it to be like this, it was essential that the body have what we now call the 'pulmonary transit' of the blood. As Servetus believed he was describing the true nature and function of respiration, an understanding which had been lost for centuries since the adoption of belief in the Trinity, we can assume that he believed that people before the Council of Nicaea had seen the body like this and that he was thus reviving an ancient view of the body, though he does not credit it to any particular ancient anatomist. What body does Servetus see? In that he is a pantheist (or perhaps even an emanationist) Servetus sees the body of man as God Himself present – perhaps as an emanation of God.[108] And he also sees the body of man as something in the course of being deified, of becoming God.

As was said at the beginning of this section, only victory in the field can decide what is or is not orthodox, who is or is not a heretic. This rule operates just as much in anatomy as in theology and thus determines which, if any, body is visible at any given time.[109] The radicals of the Reformation lost the fight, so they lost the argument. Their spiritualistic views of Christianity and the true Christian life thus turned out to be heresies. The radical views of anatomy that their religious views entailed therefore also lost, and thus did not become true views of the body. Hence the body that continued to be visible, the account which continued to be orthodox, was the Galenic one or the Galenic one reformed by Vesalius.

(iv) Counter-Reformation Rome

The relation of anatomizing to the 'Counter-Reformation', or the 'Catholic Reformation' as it is sometimes called, is not easy to elicit, at least in the example we used in Part I, that of Columbus as he turned to the anatomizing tradition of Herophilus and Erasistratus.

There are three problems. The first lies in how one is to define the Counter-Reformation and find characteristics typical of it. The 'Counter-Reformation' was not a term contemporary with the events it is now used to describe, and it is not capable of such close definition as are the terms Reformation and Radical Reformation.[110] The concept of a Catholic *Counter*-Reformation only makes sense in the light of the concept of a Protestant *Reformation* having first taken place. This has inevitably meant that the Counter-Reformation has been defined by historians primarily according to two criteria predicated on their understanding of the nature of the Protestant Reformation. First, as a *negative reaction* to Protestant innovation and informed criticism; much can be found (especially by Protestant historians) to show how *poorly* or repressively the Catholic Church responded to the northern Protestant cries for a priesthood of all believers and the reform of abuses. The second criterion used by historians to define the Counter-Reformation has been the extent to which the Counter-Reformation was *like* the Reformation; how *well* the Catholic Church responded to Protestant criticisms, the extent to which it was reforming itself all along, the presence of enlightened men within the Church, the sanctity of the lives of certain popes and so on, and it is Catholic historians, naturally enough, who have been most active in this exploration. The term 'Counter-Reformation' does not therefore embrace a readily specified series of actions or set of attitudes within the sixteenth-century Roman Catholic Church and which were identifiable *at the time* as related by their coherence and interconnection.

The second problem is that it is unclear where Columbus was when he conceived the idea of going back to the Alexandrian Ancients. The different states of Italy varied greatly in their religious positions at certain times; the singularity of one such position will be seen in the next section. It is therefore critical, if we believe that Columbus's religious position was constitutive of his anatomizing, that we know whether he was in Padua, substituting for Vesalius, or in Pisa, in the world of court patronage of the Medicis, or in papal Rome, the heart of the Counter-Reformation, when he thought that the proper anatomist ought to emulate the vivisectional anatomizing of ancient Alexandria. Columbus's pupil, Valverde, was definitely with him in Pisa, but it is not clear whether he followed him to Rome, and as Valverde was acquainted with Columbus's views on the 'new use of the lungs' he may have learnt this from the master in Pisa.

The third problem follows from this: there is nothing self-evidently 'Counter-Reformation' about Columbus's anatomizing procedure, either directly or by homological resemblance. It does not proclaim itself as Roman Catholic or even Christian (as had the approach of Paracelsus, as we saw earlier); nor does it seem to replicate in the world of anatomizing any of the reforming (or stubbornly non-reforming) activities of the Catholic Church. One can of course propose parallels at the level of metaphor to link Columbus's view of the body with his position as a Catholic anatomist, especially with respect to seeing the brain as the highest organ, literally the nerve-centre of the body, for the parallel with papal power is very attractive, but unfortunately is also inconclusive.

Can we say anything at all about Columbus and his religious position in Counter-Reformation Rome? Here we can glean something if we work by inference, particularly with respect to the known views of his 'very particular friend' Michaelangelo, whose personal physician he was.

Among the initiatives, some papal and some private and spontaneous, for the reform of the governance of the Church and of the curia, and for spiritual improvement, there was the movement associated with the name of the 'Oratory of Divine Love', whose 60 or so lay and clerical members combined together informally from about 1517 at Rome and later also in Venice and elsewhere, for the promotion of greater personal spirituality and the reform of abuses within the Church. They included a number of men later prominent in the hierarchy of the Church: Caraffa, a future pope (Paul IV); and Sadoleto, Contarini and Reginald Pole, all future cardinals. The celebrated religious reformer and poetess Vittoria Colonna was in close sympathy with them. Pope Paul III indeed appointed a commission made up of such reformers to report on the state of the Church. Its report 'Advice of select cardinals and other prelates, on emending the Church' (known as *De Emendanda Ecclesia*) was delivered in 1537.[111] Caraffa, as Pope Paul IV from 1555, turned from accommodation to a strong defence of orthodoxy. Not only did he not implement *De Emendanda Ecclesia*, which he had helped write, but he placed it on the new 'Index of prohibited books' which the Council of Trent had established in 1546. The Roman Inquisition (refounded on the model of the Spanish Inquisition of the 1470s) which had been under Caraffa's charge since 1542, was given the central role in seeking out and bringing to judgement deviants from Catholicism.

Michelangelo's religious position seems to have been of the mild reforming type associated with the 'Oratory of Divine Love'. His pupil and biographer, Condivi, stresses the point that Michelangelo was a solitary and had few friends. But among those friends, Condivi himself notes, were some of the prominent figures of the Roman 'Oratory': Reginald Pole, Claudio Tolomei and others. Michelangelo was also an intimate friend of the poetess and religious reformer, Vittoria Colonna.[112] Vasari claims that 'being a devout Christian Michelangelo loved reading the Holy Scriptures, and he held in great veneration the works written by Fra Girolamo Savonarola', the erstwhile religious reformer of Florence.[113]

From this we may be justified in assuming that Columbus shared with his 'very particular friend' views on the state of religious affairs: that he was an 'Oratory'-type reforming Catholic, and probably hostile to Protestantism. From his actions we can see also that Columbus found Counter-Reformation Rome a congenial place to live and work. For, just as Michelangelo himself did, Columbus left Pisa and the patronage and protection of the Medici court at Florence, in favour of Rome. Although Columbus first went to Rome to work with Michelangelo, he stayed for more than ten years there even though, as we saw earlier, it was clear from about 1551 that his prospective partner was no longer going to pursue their projected joint project. Columbus stayed, indeed, in the papal university itself, the *Sapienza*, which had been refounded by Pope Leo X in 1513. In other words we can assume that Columbus, living and working in the very heart of the reforming Catholic Church, had 'Counter-Reformation' attitudes. What is more, his anatomy book was dedicated to the pope, and was submitted for the *imprimatur* of the Inquisition.

However, although we can place Columbus's religious position with some precision, it does not seem possible to characterize those views in such a way as to enable us to see them as informing and shaping his reforming zeal in anatomizing, and turning him to emulate the Alexandrian Ancients. Perhaps further evidence, or evidence of a different type, may turn up in time.

(v) Reviving Aristotelian Anatomy

With respect to the revival of Aristotelian anatomizing, by contrast, the connection between religious persuasion and precise type of anatomizing is particularly strong.

In Part I, Chapter 6, I showed how Fabricius was following at Padua a programme of anatomizing based on the works of Aristotle and modelled on the anatomical practice of Aristotle. In his anatomizing Fabricius therefore took as his object of study not man, but 'The Animal', and he was seeking knowledge of the nature and functioning of the soul of 'The Animal' in all the species of animal that he could lay his hands on. I argued that the way in which Fabricius took Aristotle as his model in anatomizing exactly paralleled the way in which the professors of philosophy at Padua University also took Aristotle as their model in their philosophizing, which was equally innovative. As Zabarella said, 'since our aim is to be skilled in the philosophy of Aristotle, we must see in the first place what mode of philosophizing Aristotle himself used, and then find what we need to do so that we may be said to philosophise correctly and "aristotellically" on Aristotle'.

Why should the rulers of Venice (which controlled the University of Padua) have kept Aristotle, of all people, of all texts, at the centre of the curriculum, such that it could be read afresh now? In sixteenth-century conditions, a deliberate return to the philosophy of Aristotle is hardly the programme of radical innovators; it is the opposite of innovatory. One might almost call it reactionary. Why were the views of Venetians particularly reactionary-innovative in the sixteenth century, and hence why should they have been particularly devoted to the resuscitation of Aristotle? It had everything to do with religion, politics and trade.

For Venice, and the landward area it governed, the Veneto, was the only important place in Europe which was relatively able to hold itself free of the spiritual and political turmoils associated with the Reformation and Counter-Reformation. Venice was one of the richest states of the period by virtue of its extensive trade. It was so rich and powerful that it had taken the combination of most of the Christian princes – the pope, the emperor, the king of France, the king of Aragon, the duke of Ferrara and the marquis of Mantua – to defeat it in the war of the League of Cambrai in 1509–10. This defeat was one which Venice remembered, and her policy during the rest of the sixteenth century was shaped around the goal of keeping herself free from interference.

Venice was 'La Serenissima': the most serene republic. Its serenity consisted in the fact that it enjoyed an extraordinary stability and continuity of government; its constitution had not changed for hundreds of years, and it had never been taken over by another power. During the sixteenth century there was an active campaign by certain Venetians to celebrate this view of Venice, to celebrate the stable nature of Venetian political institutions – and hence their innate superiority to any other forms. Gasparo Contarini, for instance, wrote in the 1520s an account *Of the Magistrates and Republic of the Venetians*, which was published in 1543, celebrating the fact that

> from the first building thereof, even until this time, being now a thousand and one hundred years, it hath preserved itself free and untouched from the violence of any enemy, though being most opulent and furnished as well of gold and silver as of all other things that might, yea even from the farthest parts of the world, allure the Barbarians to so rich a booty and spoil.[114]

No state has ever lasted so long. Hence the original founders of Venice must have deliberately arranged its political institutions in such a way as to promote the common good and the common enjoyment and practice of 'virtue', and also so as to suppress the promotion of private glory. 'I dare affirm, that in the discourses of those great Philosophers, which fashioned and forged commonwealths according to the desires of the mind, there is not any to be found so well fained and framed.'[115]

The superiority of Venice comes from the fact that it is a republic under law:

> I cannot well resolve whether mankind be more beholden for any other thing to Nature, the mother of all things, than it is for this invention of laws, which was of the ancients (and not unworthily) consecrated to the immortal Gods: but which serveth more marvellous, *Aristotle* prince of philosophers, in that book of the world which he did dedicate to great *Alexander*, found not any thing to which he might likelier resemble God, than to an authentic law in a City rightly governed.[116]

Contarini claims that

> the best philosophers, and those that have learnedliest written of the ordering of a commonwealth, judged that the government thereof should be a temperature [= a mean] between the state of nobility and the popular sort, to the end that the inconveniencies of either government alone might be avoided, and the commodities of both jointly joined, for which cause they do highly commend the commonwealth of the *Lacedemonians*.[117]

Contarini claims that Venice, with its Doge and Great Council, its Senate and its Council of Ten, is precisely such a balanced constitution; hence it tends neither to tyranny nor to mob-rule. It acquired this constitution from its (mythical) founders who deliberately set out to create a state 'in accordance with Nature' and following the precepts of the philosophers; for instance,

> *Aristotle* in his *Politics* most wisely saith, that in every commonwealth which would emulate and follow the wisdom and policy of Nature, old men should be placed at the helm, and the office of the young men should be to obey and to execute those things which the old men should command them. Neither can the young men, seeing this institution is nearest unto Nature, by any means find fault or repine at the government of old men, or stir up therefore any sedition in the commonwealth.[118]

This is just how things are done at Venice. The virtues of the balanced (or 'mixed') constitution had been put forward by Aristotle in his *Politics*, and Aristotle is virtually the only authority whom Contarini cites, referring repeatedly to his *Politics* and *Ethics*. In Contarini's account, therefore, Venice is a state which emulates Nature, and accords with the precepts of Aristotle for a stable state.

The point of Contarini's account and of other similar ones, was to encourage his fellow Venetians to defend their autonomy against their sixteenth century enemies. This was the purpose that this 'myth' of the stability of Venice was meant to serve in the sixteenth-century, the so-called Golden Age of Venice, its most flourishing period, and the century which turned out to be the last one in which Venice was a world power.[119] Among the enemies against whom Venice should be protecting herself, according to its propagandists, was the Holy Roman Emperor, but by far the most important enemy in the eyes of the Venetians in the sixteenth century was the pope. It was the pope who had coordinated the 1509 League against Venice and put Venice under an Interdict. Thereafter it was the pope, of course, who wished to extend the campaign against heresy and the reform of doctrine and discipline to the whole Catholic world, including the Venetian empire. The Venetians saw this as a personal affront, an attack on their autonomy, and they actively opposed the pope. As they were such a rich and powerful trading nation they could afford to do so. They had enjoyed a long tradition of relative autonomy from papal control, even to the extent of appointing their own bishops and clergy. But it was the defence of their trade, and the independence they needed to continue it, which was the underlying motivation which led the Venetians to leap to the defence of their religious autonomy now. They needed to be able to trade with Germans and Greeks, with heretics and infidels, as well as with good Catholics. Hence they came to see a distinction between Roman Catholicism and true, Venetian, Catholicism, and they criticized the pope not for his claims to spiritual authority but to temporal authority. The pressure on Venice from the papacy continued throughout the sixteenth century, with the Venetians constantly resisting papal claims to temporal power, but allowing papal claims to spiritual authority. Hence that important instrument of the papal Counter-Reformation, the Inquisition, was indeed set up in Venice, but only after much delay, only in a form which had three Venetian laymen on the committee and with the Venetian authorities in charge of book-burning and punishment, and acting only as fervently as suited the varying needs of the Venetian authorities.[120] Heresy in general was dealt with as a state matter in Venice, rather than as an ecclesiastical one. The conflict with the pope culminated in 1606–09, when Venice was again laid under a papal Interdict. The Venetian friar Paolo Sarpi was appointed as theologian to the republic, *consultore teologico-canonico*, to counter all the arguments from Rome. He later also wrote a hostile *History of the Council of Trent*, and other works attacking the papacy's claims to temporal power.[121]

This hostility of the Venetians to the papacy should not be taken to imply that the Venetians were poor Catholics or in any sense hypocritical in their Catholicism; indeed the sixteenth century was a time of great popular piety in Venice, as the vogue for creating chapels of the holy sacrament indicates.[122] But nevertheless for most of the sixteenth century Venice was the easiest place to be of non-orthodox religious persuasion. Lutherans and other Protestants could find it a safe haven, and so on occasion could even Anabaptists. The Venetians in general were not crypto-Protestants (though of course one or two may have been), and indeed they tolerated the presence of people of other religions only so long as they did not cause trouble by trying to convert others to their beliefs. But if anywhere was a place for 'free-thinking' (the concept is not a sixteenth-century one), it was Venice. Here one was least constrained by the rival claims of Roman Catholicism and Protestantism, because the rulers of Venice – the Venetian aristocracy – were determined that the practice of religion should be under their own control and not under that of a foreign power, the papacy.[123]

What was true of Venice was true of Padua. Padua had been under Venetian rule since 1406 and its university had been treated as the Venetian university since 1434.[124] In the war of 1509 Padua had been temporarily lost to the emperor, but the Paduans had 'great love and affection' for Venetian rule and quickly opted to return to it,[125] as did other cities of the Veneto: they all found Venetian rule an easy yoke and one they preferred to all others. The university of Padua was re-opened in 1517. There was no particular conflict between Paduans and Venetians over the fact that from 1528 onwards Padua University was run from Venice by three *Riformatori dello studio* in the interests of all the citizens of the Veneto and in a way which would attract paying students from all over Europe. So at Padua, as at Venice, Venetian ideals in politics were promoted, tolerance was sustained within and toward a very large student body of different religious persuasions,[126] and, most important for our purposes, a renewed devotion was nurtured toward the Master of those who know, Aristotle.

The significance of this politico-religious situation of Venice for our present concerns is thus that in Venice and its dependent territories neither the Lutheran Reformation, nor the Spiritualist Reformation, nor the Catholic Counter-Reformation prevailed. Neither the Word of the Lutherans, nor the Spirit of the Spiritualists, nor the 'unwritten traditions' of the Roman Catholic Church was pre-eminent, and thus formative of ways of thinking in the realms of philosophy – and hence in anatomy. What prevailed instead was an old-fashioned commitment to a traditional curriculum (consisting therefore almost entirely of Aristotle), but a commitment to pursuing it without worrying overmuch about reconciling Aristotle's doctrines with those of any of the rival Christian churches. And this dual commitment, this innovation by reaction, is the key to the Venetian and Paduan transformation of Aristotelian learning in the sixteenth century. Venice was conservative, especially in the cause of preserving the best traditions for trade, but it was also more free in every sense than any other state in the sixteenth century, and the Venetian rulers were committed to a peculiarly local

Catholicism, a Venetian Catholicism rather than a Roman one. The conflict with the papacy meant that in the sixteenth century the Venetians were almost aggressive in their attempts to maintain their serenity.

The particular approach to Aristotle was unique to Padua and was promoted by the *Riformatori* who ran the university, and it complemented and was the counterpart of their specifically Venetian political and religious outlooks as members of the Venetian ruling class. So we have a situation in which one of the richest states of the period, the one where a given individual's religious ideology mattered least, and which in that sense was most 'free thinking', was promoting the teaching, in as authentic a way as possible, of Aristotle – whom today we might think of as the most reactionary and non-progressive of ancient thinkers. They and the many students they attracted were obviously finding this Aristotle liberating, fascinating and, as we have seen, finding him to be someone who opened doors to new research.

Aristotle's works were still the staple of the university curriculum everywhere in the mid-sixteenth century, except among the most highly radical protestant sects. Aristotle's texts were restored to the Lutheran course of studies after having been temporarily thrown out in the 1520s. They were fundamental, in a particular Thomist and scholastic reading, to the many new Counter-Reformation schools of the Jesuits now being established everywhere; they were taught by new Calvinists and were still being taught by old Catholics. And as virtually every university had *On the Soul* in its curriculum, the issues it raises were discussed everywhere.

But Aristotle could be many different things, and each of his books could mean something different depending on who was the interpreter. And at Padua *On the Soul* had an especial importance because there it could be discussed more freely and passionately than elsewhere, since the exercise of freedom to discuss soul matters without tying the possible answers to particular religious creeds was of such great significance to the Venetians in their continuing battle with the pope. To allow free discussion of the soul was to assert Venice's religious and political autonomy. Free discussion of the soul at Padua was important to *Riformatori*, professors and students. So important indeed that there is a story that in the 1550s a professor who was beginning to lecture on Aristotle's book *On Meteors* was shouted down by students, who cried 'We want the soul!'[127] The conclusions that the Paduan philosophers came to in their reading of Aristotle were simple but radical. Simple in that they sought to conform to what Aristotle said, radical in that they rejected a whole tradition of interpretation, and they make no particular concession to the Church's teaching on the nature of the soul, and especially on its immortality.

Most historians who have studied these Paduan philosophers agree that in their commentaries Aristotle reigns supreme, and this means that they have little to say about God. For instance, Charles Schmitt claims of Zabarella and Cremonini that:

> For these writers God seldom comes into the story. Zabarella's commentary on the *Physics* and his collection of texts called *De rebus naturalibus* ... [contain] remarkably little consideration of revealed knowledge or of God's relation to the world other than in the terms framed by Aristotle himself ... [Zabarella] was little concerned with the theological implications of philosophy.[128]

The positions held by the philosophers of Padua were not bizarre positions for people reading their Aristotle 'aristotelically', but they were found highly provocative outside the Veneto. For if you spoke like Aristotle then, whether you wished it or not, you ran the risk of being understood to deny the immortality of the soul, the creation of the universe (for Aristotle's universe was eternal), and the creative activity of God. This was potentially 'atheistic' in the sixteenth-century sense of denying that there is divine punishment or an afterlife (rather than that there is no God).[129] Cremonini was regarded by the Jesuits and the Inquisition as an enemy, and they held him under investigation for many years, but as a Paduan professor he could afford to be relatively blithe about it. It is said that rumours circulated about epitaphs he had written for himself, one of which ran 'Here lies the whole of Cremonini'[130] – meaning that there is no immortal soul – and another said, 'I was called to Philosophy and have lived entirely in it. If indeed I have sinned in philosophising, remember that I am a man, for whom it is innate to sin, but You are God, whose nature is to take pity on and always spare sinners.'[131] This sort of thing sounded very suspicious to people outside the Veneto – especially those in Rome – and the Veneto was the only place one might even hint at it as a thing possible to believe.[132] Although Cremonini is nowadays usually recalled as the Aristotelian professor who refused to look through Galileo's new telescope because it might give him a headache, he is also the one who warned Galileo not to leave Padua because he would render himself vulnerable to ecclesiastical persecution and would long for Paduan liberty again.[133]

The Venetian Senate, having sought out such outstanding professors in the first place for its university, continued to defend vigorously their right to put forward such teaching when the professors were challenged by the Inquisition. The Veneto in general was a place where relatively untrammeled discussion of such issues was possible. The *Wunderkind* of Venice, friar Paulo Sarpi, regularly attended the meetings of more than one such group of intellectuals in Venice and Padua. At the house of the Morosini in Venice there were regular meetings during the 1580s, the period when the '*Giovani*', the 'reactionary youngsters' as one might call them, were dominant in the government of Venice.[134] At these meetings there was no artificiality of manners, but

> a civil and free confidence was there only in use. It was allowed every man to make his discourse of whatsoever pleased him best, without restriction of passing from one subject to another, provided it were always of new matter, and the end of their disputation was for nothing else but to find out truth.[135]

They met not only in Venice, but in Padua too, 'whither they often transported themselves'; here they gathered in the house of Vincenzo Pinelli 'which was the receptacle of the Muses, and an Academy of all the virtues in those times'.[136]

Sarpi, the official defender of Venice against the pope, was the centre of these circles, and one of his close friends was the man who resurrected Aristotle's research programme in anatomy: Fabricius ab Aquapendente. Fabricius was 'an old friend and admirer of the father's [that is, friar Sarpi's] virtues'.[137] Now, the conflict between Venice and the pope was epitomized by the dagger which was thrust through the cheek of Sarpi one night in Venice in 1607. Everyone in Venice believed the assassins had been hired by the pope. If it was indeed the pope who was responsible for putting the knife into the Venetian patriot Sarpi, then it is significant for our story that it was Fabricius who was responsible for taking it out (see Figure 8.7). He was summoned specially from Padua by the Venetian Senate, and was made a Knight of St Mark for the successful cure of Sarpi. Fabricius attended Sarpi medically in other ways, and constructed a special syringe or catheter for Sarpi, who often suffered from suppression of urine.[138] Sarpi had as a young man (probably in the early 1580s) 'cut in pieces a number of living creatures with his own hands to make Anatomies', and much of this work seems to have been done in association with Fabricius. Indeed Sarpi's admiring biographer hints that Fabricius virtually stole from Sarpi various anatomical discoveries, including that most famous one of the *ostiola* in the veins.[139] Sarpi has even been credited with having designed the 'Fabrician' anatomical theatre at Padua.[140]

With respect to his religious position, we do not need to assume that Fabricius was an atheist any more than his colleagues were; rather, the likelihood is that he was a Venetian Catholic. He speaks often in praise of God the Creator in his writings. And given the concentration in his anatomical work on finding intelligence, purpose and design in Nature, it may be claimed that Fabricius's anatomizing is even more God-oriented and even less medically-oriented than that of any of the Moderns in anatomy that we have dealt with so far.

8.7 Portrait of Sarpi, 'eviscerator of the Council of Trent'. The scar of the dagger wound in his cheek, acquired in the defence of the liberty of Venice, is clearly visible.

So we can say that by association, as well as by his professional role as Professor of Anatomy and Surgery at Padua, Fabricius was someone who was a full member of the Venetian intellectual elite, and hence that, like other members of that group, he would be keen to read Aristotle's anatomical works 'aristotelically', and to discuss them without reference to their orthodoxy (or otherwise) in the eyes of the pope. We may even say that it was his moral duty, as a Venetian anatomist, to study anatomy aristotelically. Indeed, although it has not previously been appreciated, the best evidence we have of the very strong support that the Venetian authorities and the *Riformatori* gave to reading Aristotle freely and aristotelically is to be found precisely in the career and work of Fabricius himself, as we have seen.

(vi) Conclusions

'Why?' questions are never easy to answer in history. I have argued that there was a sequence of anatomists turning back to progressively more ancient Ancients in the sixteenth century and imitating their projects in anatomy, and I hope I have demonstrated this beyond doubt. I have also argued that anatomizing in the sixteenth century and the centuries around it was part of a now-defunct discipline, Natural Philosophy, which had the admiration of God's creative providence at its core. Further, I have laid out the arguments (some of them built on the work of other historians) that the Renaissance is a category empty of historical explanatory power. I turned, instead, to the Reformation as a potentially better explanatory category for the dynamic changes in anatomical practice in the sixteenth century. What success has this given us in exploring why particular Modern anatomists turned to particular Ancient anatomists and sought to emulate their practice?

In some cases the explanatory power of the Reformation, and reactions to it, seems to be very strong and to account for why particular anatomists in particular local circumstances adopted certain attitudes to anatomizing and/or took up certain ancient anatomical practices. Given the significance of Aristotle's philosophy for the Venetians in resisting the temporal claims of the papacy, then the model of Aristotle's anatomizing was the obvious one for Fabricius, a member of the Venetian intellectual elite and personal friend of its official defender, to adopt and put into practice. Aristotelian anatomy was almost a patriotic duty for Fabricius to pursue, and it was also fruitful and valuable in its own right, giving one new understanding about anatomy and the actions of the soul. The case is equally strong as to why Paracelsus, as a Spiritualist physician, disdained the Greek anatomical tradition completely, rejecting it for a Christian and inspired approach to understanding how the body functions in health and disease. Similarly in the case of Servetus we can see how someone trained in the Greek anatomical tradition, yet subscribing to an extreme pantheistic Christianity in which the Holy Spirit is breathed in with every breath, could find new anatomical pathways in the body which must exist to allow the Holy Spirit to come into the body so that the Substance of God can be mingled with the Substance of man.

In the case of Paracelsus and Servetus their language is religious as well as medical or anatomical; both of them have strong religious agendas. With Fabricius one can see him working integrally within a politico-religious agenda of a very distinctive state, and hence being led, quite naturally, to undertake anatomy in a certain way: the religious confession is only one step away from the anatomy.

But in the other cases, the connection is not so strong. I argued that with respect to Vesalius the pattern of his behaviour and certain very specific aspects of his ideology and behaviour in anatomizing are exactly the equivalents of Luther's in religious reform, and that therefore we might be within our rights in claiming that what motivated him to undertake his anatomical reforms was a pre-commitment to Lutheran religious ideals. But without firm evidence of Vesalius's religious commitment from some other source, this can only be an inference. With Sylvius and Guenther and with Columbus, our information simply is not adequate at the moment even to make realistic inferences about their religious commitments and hence about the motivation for their particular forms of anatomizing, even though we can find interesting and suggestive parallels with religious reform in the first of these cases.

But our inability, at least at present, to make watertight cases about the relation of religious confession to anatomizing agenda, should not therefore lead us to revert to the old assumption that religion and 'science' (as represented by anatomy) were discrete and unrelated fields in the sixteenth century. Anatomy showed God in action; it was not a secular activity nor one which led to a secularizing world-view. The soul was all-important: as the medical students shouted at Padua in the 1550s, 'We want the soul!'

Notes

1. Frederic Seebohm, *The Oxford Reformers of 1498: A History of the Fellow-work of John Colet, Erasmus, and Thomas More* (first published 1867), London, Dent, 1938, p. 52.
2. Margaret Mann Phillips, *Erasmus and the Northern Renaissance*, London, Hodder and Stoughton, 1949, p. 46.
3. My translation, from the (unpaginated) 'Lectori'.
4. *An Exhortation to the Diligent Studye of Scripture Made by Erasmus of Roterodamus. And Translated into Inglish*, London, 1529; the Cambridge University Library copy (spelling modernized).
5. Marjorie O'Rorke Boyle, *Christening Pagan Mysteries: Erasmus in Pursuit of Wisdom*, University of Toronto Press, 1981; on the pagan mysteries, see Edgar Wind, Pagan *Mysteries in the Renaissance*, Oxford University Press, 1980 (first edn 1958).
6. My translation from *Omnia Platonis Opera*, Venice, Aldus, 1513, f. 3r.

Much of the dedication by Aldus (but not the part translated above) and the whole of the Greek dedicatory verse by Musurus, are translated in William Roscoe, *The Life and Pontificate of Leo the Tenth*, Liverpool, 4 vols, 1805, vol. 2, pp. 241–52.

7. Of the many studies of Luther E.G. Schwiebert, *Luther and his Times: The Reformation from a New Perspective*, Saint Louis, Missouri, Concordia, 1950, is still very valuable. The General Introduction to *Luther: Lectures on Romans*, translated and edited by Wilhelm Pauck, Philadelphia, Westminster Press, 1961, is highly informative on Luther's methods. On the state of Wittenberg see M. Grossmann, *Humanism in Wittenberg 1485–1717*, Nieuwkoop, 1975. For Luther's works in the original Latin or German see *D. Martin Luthers Werke* (various editors), Weimar, 1883–present; for translations of them into English see *Luther's Works*, ed. by Jaroslav Pelikan and Helmut T. Lehmann, Philadelphia, 55 vols, 1955–.
8. On Melanchthon see Clyde L. Manschreck, *Melanchthon: The Quiet Reformer*, New York, Abingdon Press, 1958; Gerald Strauss, *Luther's House of Learning: Indoctrination of the Young in the German Reformation*, Baltimore, Johns Hopkins University Press, 1978.
9. Carl C. Christensen, *Art and the Reformation in Germany*, Ohio University Press, Wayne State University Press, 1979. For treatment of other aspects of Lutheran and Reformation iconography, see R.W. Scribner, *For the Sake of Simple Folk: Popular Propaganda for the German Reformation*, Cambridge University Press, 1981, and *Popular Culture and Popular Movements in Reformation Germany*, London, Hambledon, 1987.
10. Christensen, op.cit., p. 110.
11. Ibid., p. 150. This comes from Luther's 1530 *Commentary on Psalm 111*; it is translated in Luther's Works vol. 13, p. 375, and the original German is in *Werke*, vol. 31i, p. 415. The verses in golden letters read: 'Der Gnedige und Barmherziger HERR hat ein gedechtnis seiner wunder gestifft'.
12. Compare the tradition in which, e.g. popes had themselves portrayed 'as' St Jerome or even St Peter; this (I believe) is about representing the person portrayed either in their office (St Peter, head of the church), or representing them as having, by attribution, the qualities or virtues of the saint in question (e.g. the learnedness of Jerome). The case described above, however, is about Luther and his allies being portrayed as *being* the Apostles themselves, reliving the apostolic story. On Luther believing himself to be St Paul, see M.V. Edwards Jr, *Luther and the False Brethren*, Stanford University Press, 1975.
13. *Luther's Works*, vol. 31, pp. 345–46.
14. *Luther's Works*, vol. 51, p. 77.
15. Erik Erikson, *Young Man Luther: a Study in Psychoanalysis and History*, London, Faber, 1972 (first published 1959).
16. It was Cajetan, the General of the Dominicans, who ordered Luther to keep silence, in October 1518; the Dominicans, of course, were traditionally the Inquisitors.
17. H.R. Trevor-Roper, *Religion, the Reformation and Social Change*, London, Macmillan, 1967, p. 24.
18. See Henri Pirenne, *Histoire de Belgique*, 7 vols, Brussels, 1902–32, vol. 3, ch. 4; the first martyrs were burnt in Brussels in 1523.

19. See David S. Hempsall, 'Martin Luther and the Sorbonne, 1519–21', *Bulletin of the Institute for Historical Research* 1973, **46**, pp. 28–40, and his 'Measures to Suppress "La Peste Luthérienne" in France, 1521–22', in the same journal, 1976, **49**, pp. 296–99; Victor Carrière, 'La Sorbonne et l'Evangelisme au XVIe siècle', in L. Halphen (ed.) *Aspects de l'Université de Paris*, Paris, 1950, pp. 159–86; Walter F. Bense, 'Noel Beda and the Humanist Reformation at Paris, 1504–1534', Unpublished PhD Thesis, Harvard University, 1967.
20. See C.D. O'Malley, *Andreas Vesalius of Brussels, 1514–1564*, University of California Press, 1964, p. 203 for Vesalius refusing a post at Pisa in 1544.
21. *Epistola, Docens Venam Axillarem Dextri Cubiti In Dolore Laterali Secandam: Et Melancholicum Succum Ex Venae Portae Ramis Ad Sedem Pertinentibus, Purgari*, Basle, Winter, p. 30. For a translation of this see John B. deC.M. Saunders and Charles Donald O'Malley, 'Andreas Vesalius Bruxellensis, The Bloodletting Letter of 1539: An Annotated Translation and Study of the Evolution of Vesalius's Scientific Development', in *Studies and Essays in the History of Science and Learning Offered in Homage to George Sarton on the Occasion of his Sixtieth Birthday 31 August 1944*, ed. by M.F. Ashley Montague), New York, Schuman, 1946; pp. 48–49.
22. This is as translated by O'Malley op.cit., p. 111. The Latin is as follows: 'Ego ex me vel inde conjecturam facere possum, quod Galeni librum de Ossibus vel ter studiosis praelegerim, priusquam mendam aliquam Galeni annotare ausus fuissem: quum tamen nunc satis nequeam mirari meam stupiditatem, qui, quae scribebantur, tam parum assequerer, et ipse meis oculis ita imponerem', *Andreae Vesalii Bruxellenisis, Medici Caesarei Epistola, Rationem Modumque Propinandi Radicis Chynae Decocti . . . Pertractans*, Basle, 1566, p. 196.
23. See Saunders and O'Malley, op.cit., esp. pp. 66–68; Vesalius's language is quite striking, and he raises the issue several times.
24. Eriksson op.cit., p. 272–73.
25. Eriksson op.cit., pp. 290–91.
26. Eriksson op.cit., pp. 292–93; Vesalius gave his answer at the end of the previous demonstration: viz., that the two movements are different in time.
27. On the willingness of Oporinus, Vesalius's printer, to print unorthodox works, see O'Malley, op.cit., pp. 132–33.
28. Cushing, p. xxviii; on the pilgrimage see also G.M. Cullen, 'The Passing of Vesalius', *Edinburgh Medical Journal* 1918, **13**, pp. 324–39, and 388–440, and 'Vesalius and the Inquisition Myth', *Lancet*, 1929, 2, p. 362.
29. For this dating see C.G. Bretschneider, *Corpus Reformatorum: Opera Melanchthonis*, Halle, vol. 13, p.1.
30. This is in fact the opposite of Aristotle's own sequence. It is clear from *De Anima* that Aristotle saw the work as preliminary to the accounts preserved in the animal books.
31. The occasion was the problems with the Anabaptists. On Melanchthon's anatomy see Vivian Nutton, 'Wittenberg anatomy' in Ole Peter Grell and Andrew Cunningham, eds, *Medicine and the Reformation*, London, 1993, pp. 11–32; Nutton claims that this anatomy teaching was intro-

duced because of the particular importance for Protestants of the relationship between body and soul in revealing the workings of God and promoting Christian morality. On Melanchthon's view of the soul, see Vivian Nutton, '*De Placitis Hippocratis et Platonis* in the Renaissance', in P. Manuli and M. Vegetti, *Le Opere Psicologiche di Galeno: Atti del Terzo Colloquio Galenico Internazionale, Pavia, 1986*, Bibliopolis, 1988, pp. 281–309, esp. 303–5.

32. f. a 1v – a 2r.
33. f. 44 v – 45 r; my translation.
34. This has been reprinted in *Corpus Reformatorum* vol. 13, and its dedicatory letter in vol. 7, cols 1123–28.
35. Melanchthon's copy of the *Fabrica* is now in the Surgeon General's Library; I am most grateful to the Librarian for supplying me with a xerox of the poem. The poem is translated in Cushing, op.cit., pp. 80–81, but he does not print the Latin. Melanchthon wrote this poem at Nuremberg where he was helping found a school; that is to say, in an immediate teaching context. The translation into English verse used above is by Dorothy Schullian, 'Old volumes Shake their Vellum Heads', *Bulletin of the Medical Library Association*, 1945, **4**, pp. 440.
36. Ibid., vol. 7, col. 1127; my translation. For other compliments to Vesalius see the poem by Johannes Stigelius printed in the original edition (but not reprinted in *Corpus Reformatorum*, vol. 13); and *Corpus Reformatorum* vol. 13, cols 21, 31 and 62. Fuchs of course was a Lutheran.
37. There were also two printings of Vesalius's *Epitome* at Wittenberg, in 1582 and 1603. For references on anatomy in other reformed philosophy courses, see Andrew Cunningham, 'The Kinds of Anatomy', *Medical History*, 1975, **19**, pp. 1–19, see pp. 6–7; the work used in the Scottish universities was Fernel, not Vesalius or Melanchthon, and their reform was Calvinist rather than Lutheran.
38. Loc. cit., esp. pp. 214–20, 238–40, 246–51; on Sylvius's response see Chapter 4, above.
39. The authoritative work on this is George H. Williams, *The Radical Reformation*, Philadelphia, Westminster Press, 1962.
40. Ronald J. Sider, *Andreas Bodenstein von Karlstadt: The Development of his Thought, 1517–1525*, Leiden, Brill, 1974, pp. 92–93, quoting Karlstadt.
41. Ibid., p. 93.
42. On this see Norman Cohn, *The Pursuit of the Millenium: Revolutionary Millenarians and Mystical Anarchists of the Middle Ages*, revised and expanded edition, London, Temple Smith, 1970, pp. 234–51.
43. On the mystical dimension see Steven E. Ozment, *Mysticism and Dissent: Religious Ideology and Social Protest in the Sixteenth Century*, Yale University Press, 1973, and on the apocalyptic see Cohn, op.cit.
44. August 1524, in the *Special Exposé of False Faith*, in Ozment, op.cit.; the quotations are from pp. 80, 83, 84, 85 and 87.
45. As cited by Hans-Jurgen Goertz, 'The Mystic with the Hammer: Thomas Müntzer's Theological Basis for Revolution', in James M. Stayer and Werner O. Packull (eds), *The Anabaptists and Thomas Müntzer*, Iowa, Kendall, 1980, pp. 118–32, see p. 121.
46. Ozment, op.cit., p. 89.
47. An unidentified speaker, in 1538, cited in Harold S. Bender, 'The Signifi-

cance of Sixteenth-Century Anabaptism', in Stayer and Packull, op.cit., pp. 13–22, see p.15.

48. Cohn, op.cit., gives an excellent account of this in Chapter 13, 'The Egalitarian Millenium (iii)'.

49. Bernd Moeller, *Imperial Cities and the Reformation: Three Essays*, edited and translated by H.C. Erik Midelfort and Mark U. Edwards Jr., Philadelphia, Fortress Press, 1972, p. 53; Basil Hall, 'The Reformation City', *Bulletin of the John Rylands Library*, 1971, 54, pp. 103–48; for an example, see R.W. Scribner, 'Civic Unity and the Reformation in Erfurt', *Past and Present*, 1975, no. 66, pp. 29–60.

50. See the listing in Moeller, op.cit., p. 91.

51. Robert C. Walton, *Zwingli's Theocracy*, University of Toronto Press, 1967, p. xxii.

52. On Celsus, the late Roman writer whose book on medicine, *De medicina*, was discovered and published in 1478, see Chapter 1 above. Most important on Paracelsus in general is Walter Pagel, *Paracelsus: An Introduction to Philosophical Medicine in the Era of the Renaissance*, Basel, Karger, 1958; Charles Webster, 'Paracelsus: medicine as popular protest', in Ole Peter Grell and Andrew Cunningham, eds, *Medicine and the Reformation*, London, Routledge, 1993, pp. 57–77. The medical works of Paracelsus in the original German or Latin have been edited by Karl Sudhoff, *Theophrastus von Hohenheim, genannt Paracelsus: Sämtliche Werke*, Munich and Berlin, Oldenbourg, 1929–33, 14 vols. Some of the medical writings are accessible in A.E. Waite, *The Hermetic and Alchemical Writings of Aureolus Philippus Theophrastus Bombast, of Hohenheim, called Paracelsus the Great*, London, Elliott, 1894, 2 vols; repr. New York, University Books Inc., 1967; Jolande Jacobi, *Paracelsus: Selected Writings*, translated by Norbert Guterman, London, Routledge and Kegan Paul, 1951 (German edition 1942); Kurt F. Leidecker has translated *Volumen Medicinae Paramirum*, Supplement no. 11 to the *Bulletin of the History of Medicine*, 1949; Henry Sigerist has edited *Four Treatises*, Baltimore, Johns Hopkins University Press, 1941. On some of the problems of the interpretation of Paracelsus see Owsei Temkin, 'The Elusiveness of Paracelsus', *Bulletin of the History of Medicine*, 1952, 26, pp. 201–17.

53. Translated by George Rosen and printed in Sigerist, *Four Treatises*, pp. 43–126.

54. He has been identified as a Spiritualist by the most authoritative historian of these matters, George Williams, op.cit. pp. 192, 197. As Paracelsus died still technically a Catholic, and as his doctrines do not always coincide with those of other mid-sixteenth century 'Spiritualists', some scholars hesitate to ascribe the title to Paracelsus. For instance, unlike other Spiritualists, Paracelsus saw the eucharist as important, but 'as *pharmakon athanasius*, the imparting of the immortal heavenly medicine to the earthly and mortal, of the sacramental to the elemental body' (Hartmuth Rudolph, 'Theophrast von Hohenheim (Paracelsus): Physician and Apostle of the New Creation', in Hans-Jurgen Goertz (ed.), *Profiles of Radical Reformers*, Ontario, Herald Press, 1982, pp. 255–68, see p. 265). Given that 'pharmakon athanasius' means 'medicine conferring immortality', such a position might well be seen as entirely appropriate for a physician Spiritualist. On the difficulties of finding a consist-

ent definition of 'Spiritualist' and of identifying Spiritualists in this period, see the entry 'Spiritualism' in *The Mennonite Encyclopaedia*, Kansas, 1959. On Paracelsus's religious position see also Pagel, op.cit., pp. 40–44. The account by Oporinus, who made himself Paracelsus's amanuensis for two years and translated some of his works into Latin, is inconclusive with respect to Paracelsus's religious views, merely saying that he never prayed and that he equally disdained Catholicism, Basle evangelicism (i.e. that led by Oecolampadius), Luther and the pope; for the Latin text of Oporinus's letter see the unsympathetic account of Daniel Sennert, *De Chymicorum cum Aristotelicis et Galenicis Consensu ac Dissensu Liber*, Paris, third edition 1633, pp. 32–33; on the relations of Paracelsus and Oporinus see Von Udo Benzenhofer, 'Zum Brief des Johannes Oporinus uber Paracelsus', *Sudhoffs Archiv*, 1989, 73, pp. 55–63.

55. Williams, op.cit., p. 147.
56. Hans R. Guggisberg, *Basle in the Sixteenth Century: Aspects of the City Republic Before, During and After the Reformation*, St Louis, Missouri, Centre for Information Research, 1982, p. 24.
57. For Paracelsus's call to Basle, details of his lecturing there, and the hostility of the faculty to him, see the articles by R.-H. Blaser, reprinted in *Paracelsus in Basel: Festschrift fur Professor Dr. Robert-Henri Blaser zur 60. Geburtstag*, Basel, 1979.
58. On the meaning of *ratio* here, see below.
59. My translation from the Latin text in Sudhoff, *Sämtliche Werke*, vol. 4 (1931), pp. 3–4. Waite offers an English translation in vol. 2, pp. 169–71. Presumably Paracelsus composed this originally in German, and it was translated into its surviving Latin form by Oporinus or some other amanuensis; for Oporinus, see above note 54.
60. On this meaning of 'experience' for mystics and Spiritualists, see Ozment, *Mysticism and Dissent*, Ch. 2, *passim*. It is nicely encapsulated by Castillio, writing in 1556, 'For true knowledge consists of experience in the sense that he who has had the power to do something knows once and for all and for certain what he can do', quoted in Ozment, p. 45.
61. Jacobi, op.cit., pp. 175–76, 178, and for definitions see p. 327 (under 'Light of Nature'), and 331 (under 'Reveal, Manifest'); Walter Pagel, *The Smiling Spleen: Paracelsianism in Storm and Stress*, Basel, Karger, 1984, p. 98.
62. As quoted in Jacobi, op.cit., pp. 255–56.
63. See *Volumen Medicinae Paramirum*.
64. Pagel, Paracelsus, p. 83.
65. See 'Alchemy: the Third Column of Medicine' (from the *Liber Paragranum* of 1529–30), as translated in Waite, op.cit., vol. 2, pp. 148–64.
66. Jacobi., op.cit., pp. 158, 166–67; *Labyrinthus Medicorum*, translated in Waite, op.cit., vol. 2, pp. 165–68; P.M. Rattansi, 'Art and Science: The Paracelsian Vision', in F.D. Hoeniger and J. Shirley (eds), *Science and the Arts in the Renaissance*, New Jersey, Folger Shakespeare Library, 1985, pp. 50–58.
67. Pagel, *Paracelsus*, has explored these features extensively. On the astral body, see D.P. Walker, 'The Astral Body in Renaissance Medicine', *Journal of the Warburg and Courtauld Institutes*, 1958, **21**, pp. 119–33. Paracelsus had no problem with the fact that some of these things are

non- or pre-Christian, because he subscribed to the view that they were part of a primitive theology or *prisca theologia.*

68. This is a text Paracelsus himself cites; see for instance Jacobi, op.cit., p. 239.
69. As translated by Harris L. Coulter, *A Divided Legacy: A History of the Schism in Medical Thought*, Washington, Wehawken Book Co., 3 vols, 1973–77, vol. 1, p. 407, from the German in *Sämtliche Werke*, vol. 6, pp. 332–33.
70. My translation from Ioannis Guenther, *De Medicina Veri et Nova tum Cognoscenda, tum Faciunda Commentaria Duo*, Basle, 1571, Second Commentary, p. 28.
71. See Walter Pagel and Pyarali Rattansi, 'Vesalius and Paracelsus', *Medical History*, 1964, **8**, pp. 309–28; see esp. 310–15. See also Pagel *Paracelsus*, op.cit., esp. pp. 134–40.
72. See Charles Webster, *The Great Instauration: Science, Medicine and Reform 1626–1660*, London, Duckworth, 1975; idem, 'Alchemical and Paracelsian Medicine' in Charles Webster (ed.), *Health, Medicine and Mortality in the Sixteenth Century*, Cambridge University Press, 1979, pp. 301–34; idem, *From Paracelsus to Newton: Magic and the Making of Modern Science*, Cambridge University Press, 1982; idem, 'English Medical Reformers of the Puritan Revolution: a background study to the "Society of Chymical Physitians"', *Ambix*, 1967, **14**, pp. 16–41; P.M. Rattansi, 'Paracelsus and the Puritan Revolution', *Ambix*, 1963, **11**, pp. 24–32, and 'The Helmontian–Galenist controversy in Restoration England', *Ambix*, 1964, **12**, pp. 1–23.
73. Guenther, op.cit., is an example of this. On the creation of a non-Paracelsian, indeed anti-Paracelsian, chemistry by a Lutheran at the end of the sixteenth century, and on the creation of a Paracelsian chemical medicine based on a Cabalistic view of the Word, yet compatible with Calvinism, see Owen Hannaway, *The Chemists and the Word: the Didactic Origins of Chemistry*, Baltimore, Johns Hopkins University Press, 1975. For the career of Paracelsianism, see Allen G. Debus, *The Chemical Philosophy: Paracelsian Science and Medicine in the Sixteenth and Seventeenth Centuries*, New York, Science History Publications, 2 vols, 1977; idem, *The English Paracelsians*, London, Oldbourne, 1965.
74. The most important modern works on Servetus are Roland H. Bainton, *Hunted Heretic: The Life and Death of Michael Servetus, 1511–1553*, Boston, The Beacon Press, 1953, revised edn 1960; Earl Morse Wilbur, *A History of Unitarianism: Socinianism and its Antecedents*, Cambridge, MA, Harvard University Press, 1946; there is much important information in George H. Williams, *The Radical Reformation*, Philadelphia, The Westminster Press, 1962. See also Jerome Friedman, 'Michael Servetus' in *Profiles of Radical Reformers*, ed. by Hans-Jurgen Goertz, Ontario, Herald Press, 1982, pp. 247–54. Earl Morse Wilbur has translated *The Two Treatises of Servetus on the Trinity* (Cambridge, MA, Harvard University Press; Harvard Theological Studies, 1932), and his Introduction is very useful. For translations of Servetus's minor books see Charles Donald O'Malley, *Michael Servetus: A Translation of his Geographical, Medical and Astrological Writings*, Philadelphia, American Philosophical Society, 1953; the major work we are concerned with, *The Restitution of Christianity*, has not been translated into English as a whole (the

pertinent passage is translated by O'Malley), though the Latin text of the 1790 reprint of the Vienna copy is available in a 1966 re-reprint by Minerva G.M.B.H. of Frankfurt a. M. Much of Servetus's writing is paraphrased or translated in R. Willis, *Servetus and Calvin*, London, King, 1877. For surviving original copies of Servetus's works see John M. Fulton, *Michael Servetus: Humanist and Martyr; With a Bibliography of his Works in a Census of Known Copies by Madeline E. Stanton*, New York, Reichner, 1953. On Servetus and other themes of the present chapter, see S.F. Mason, 'The Scientific Revolution and the Protestant Reformation', *Annals of Science*, 1953, 9, pp. 64–87 and 154–75.

75. See, for instance, William Osler, *Michael Servetus*, Oxford University Press, 1909.
76. See Jerome Friedman, *Michael Servetus: A Case Study in Total Heresy*, Geneva, Libraire Droz S.A., 1978 (Travaux d'Humanisme et Renaissance No CLXIII).
77. My account here follows Bainton, op.cit. On the parallel population of moriscos in this period, forcibly converted from Islam, and their medicine, see Luis García-Ballester, 'Academicism Versus Empiricism in Practical Medicine in Sixteenth Century Spain with regard to Morisco Practitioners', in A. Wear, R.K. French and I.M. Lonie, *The Medical Renaissance of the Sixteenth Century*, Cambridge University Press, 1985, pp. 246–70; and his 'The Inquisition and Minority Medical Practitioners in Counter-Reformation Spain: Judaizing and Morisco Practitioners, 1560–1610', in Ole Peter Grell and Andrew Cunningham, eds, *Medicine and the Reformation*, London, Routledge, 1993, pp. 156–91.
78. *Two Treatises*, as translated by Morse, p. 50.
79. On which see Morse, *History of Unitarianism*, Chapter 2, 'Latent Anti-Trinitarianism among the early Reformers'.
80. Ibid., pp. 7–8.
81. Bainton, op.cit., pp. 67–68.
82. The dating of the events of Servetus's life is difficult, and has been built by scholars from the evidence Servetus gave at his various trials.
83. In the 1536 *Apology Against Fuchs*, Servetus calls himself a pupil of Champier; see O'Malley, op.cit., p. 44.
84. See ibid., p. 55, for the dating which shows that Vesalius and Servetus could not have been pupils together.
85. My translation from the 1539 Basle edition of Guenther's *Institutionum Anatomicarum Secundum Galeni Sententiam*, p. 7.
86. Bainton, op.cit., p. 145.
87. See Morse, *History of Unitarianism*, p. 168: this was 'not merely the trial of an individual for heresy, but a contest for supremacy in the government of Geneva'. See also Bainton, op.cit., pp. 193–94, and E. William Monter, *Calvin's Geneva*, New York, Wiley, 1967.
88. On which see Morse, *History of Unitarianism*, Chapters 13 and 14; Guggisberg, op.cit., Chapter 3; B. Becker, ed., *Autour de Michel Servet et de Sebastian Castellion*, Haarlem, Tjeenk Willink & Zoon N.V., 1953. Oporinus, the Basle printer, again enters our story here. He had been a follower and amanuensis to Paracelsus in 1527–28 (above, note 37), printer of Vesalius's great book in 1543, and now in 1553 and 1554 printer of the pro-Servetus attacks on Calvin by Sebastian Castellio.
89. Bainton, op.cit., p. 73.

90. *Two Treatises*, translated by Morse, pp. 33, 75–76.
91. Ibid., p. 67.
92. Ibid., pp. 90, 83.
93. On which see Morse, *History of Unitarianism*, pp. 141–43, and Friedman, *Total Heresy*, pp. 36–42.
94. Bainton, op.cit., p. 140.
95. *Two Treatises*, translated by Morse, p. 44.
96. Ibid., pp. 94, 93, 103.
97. Ibid., p. 258.
98. Ibid., p. 209.
99. As translated by Friedmann, *Total Heresy*, p. 69; and on deification see pp. 61–74.
100. On this see ibid., pp. 55–56, and 137–39 where the extent of Champier's influence is assessed. On Champier, see Brian P. Copenhaver, *Symphorien Champier and the Reception of the Occultist Tradition in Renaissance France*, Hague, Mouton, 1978; and M.P. Allut, *Etude Biographique et Bibliographique sur Symphorien Champier*, Lyons, Scheuring, 1859; and Claudio Manzoni, *Umanesimo ed Eresia: Michele Serveto*, Naples, Guida, 1974.
101. *Christianismi Restitutio*, p. 169, my translation. For a general summary of the contents of the book and translations of some pertinent passages, see Willis (note 74 above), and for the passage translated in extenso see O'Malley (note 74 above).
102. Ibid.
103. Ibid., pp. 169–70; I have deliberately chosen to translate anima as 'soul' rather than 'life', because I believe that this is what Servetus is referring to.
104. Ibid., p. 170.
105. Ibid.
106. Ibid., p. 171.
107. Ibid., pp. 170, 171.
108. On Servetus's emanationist views, see Morse, *History of Unitarianism*, pp. 144–45, Friedman, *Total Heresy*, *passim*.
109. That truth about Nature is determined by power relations between people, has been brilliantly argued by Bruno Latour, *Science in Action: How to Follow Scientists and Engineers through Society*, Milton Keynes, Open University Press, 1987, and would seem to have applied to Natural Philosophy as much as it now does to science.
110. On the Counter-Reformation in general see H. Outram Evennett, *The Spirit of the Counter-Reformation*, Cambridge University Press, 1968, who has a discussion of the difficulties of the term; G.W. Searle, *The Counter Reformation*, University of London Press, 1974; A.G. Dickens, *The Counter Reformation*, London, Thames and Hudson, 1968. A.D. Wright, *The Counter-Reformation*, London, Weidenfeld and Nicholson, 1982, offers a revisionist view of the nature and consequences of the Counter Reformation.
111. This is available in English translation in Elisabeth G. Gleason, *Reform Thought in Sixteenth Century Italy*, Chico (California), Scholars Press, 1981, pp. 81–100.
112. Ascanio Condivi, *The Life of Michelangelo*, translated by Alice Sedgwick Wohl, ed. by Hellmut Wohl, Louisiana State University Press, 1976, pp.

102–03; Giorgio Vasari, *Life of Michelangelo Buonarroti*, translated by George Bull, London, Folio Society, 1971 (from the 1568 version), pp. 122–23, 125.

113. Vasari, op.cit., 125.
114. Quoted from Gasparo Contarini, *The Commonwealth and Gouernment of Venice*, translated by Lewis Lewkenor, London, Mattes, 1599, reprinted in facsimile by Da Capo Press, Amsterdam and New York, 1969, p. 5; I have modernized the spelling in quotations from Contarini. The most extensive treatment of political defenses of Venice is William J. Bouwsma, *Venice and the Defense of Republican Liberty*, University of California Press, 1968.
115. Contarini, p. 7.
116. Ibid., pp. 11–12. The book reputed to have been dedicated to Alexander is the *Politics*.
117. Ibid., p. 14; 'the commonwealth of the *Lacedemonians*' is Sparta, whose case Aristotle refers to in the *Politics*.
118. Ibid., pp. 64–65.
119. See Bouwsma, op.cit.; J.R. Hale, ed., *Renaissance Venice*, London, Faber and Faber, 1973; Jane Martineau and Charles Hope, eds, *The Genius of Venice 1500–1600*, London, Royal Academy of Arts, 1983.
120. See P.F. Grendler, *The Roman Inquisition and the Venetian Press, 1540–1605*, Princeton University Press, 1977.
121. For a full account of Sarpi's *History* see Bouwsma, op.cit.
122. See M.E. Cope, 'The Venetian Chapel of the Sacrament in the Sixteenth Century: A Study in the Iconography of the Early Counter-Reformation', PhD dissertation, University of Chicago, 1965, published in series 'Outstanding Dissertations in the Fine Arts', New York, Garland, 1979.
123. See Bouwsma, op.cit., and for the 1540s see Antonio Santosuosso, 'Religious Orthodoxy, Dissent and Suppression in Venice in the 1540s', *Church History*, 1973, 42, pp. 476–85.
124. There was also a sort of university in Venice itself; on which see Richard Palmer, *The 'Studio' of Venice and its Graduates in the Sixteenth Century*, Contributi alla Storia dell'Universita di Padova no. 12, Padua, LINT, 1983.
125. Contarini, op.cit., p. 147.
126. See Peter J. van Kessel, 'The Denominational Pluriformity of the German Nations at Padua and the Problem of Intolerance in the 16th Century', *Archiv für Reformationsgeschichte*, 1984, 75, pp. 256–76.
127. Quoted in J.H. Randall, Jr, *The Career of Philosophy*, Columbia University Press, 1962, p. 68. On the soul in the sixteenth century, see F. Edward Cranz, 'The Renaissance Reading of "De Anima"', in *Platon et Aristote à la Renaissance*, XVIe Colloque International de Tours, Paris, Vrin, 1976, pp. 359–76; Katherine Park and Eckhard Kessler, 'The Concept of Psychology', Katherine Park, 'The Organic Soul', and Eckhard Kessler, 'The Intellective Soul', all in Quentin Skinner and Eckhard Kessler, eds, *The Cambridge History of Renaissance Philosophy*, Cambridge University Press, 1988, pp. 455–63, 464–84 and 485–534 respectively.
128. Schmitt, *Aristotle and the Renaissance*, pp. 31–32.
129. For this definition see David Wootton, *Paolo Sarpi: Between Renaissance and Enlightenment*, Cambridge University Press, 1983, p. 5.

130. 'Cremoninus hic totus jacet'; quoted in J.-Roger Charbonnel, *La Pensée Italienne au XVIe Siècle et le Courant Libertin*, Paris, Champion, 1919, p. 271, n.
131. 'Ad philosophiam sum vocatus, in eam totus fui; si aliquid philosophando peccavi, memento me esse hominem, cui innatum est peccare; te vero esse Deum cui proprium est misereri et semper parcere'; quoted in Charbonnel, op.cit., p. 272, n., my translation. On Cremonini see the still highly valuable account by Leopold Mabilleau, *Etude Historique sur la Philosophie de la Renaissance en Italie (Cesare Cremonini)*, Paris, Hachette, 1881.
132. On this issue see Paul Oskar Kristeller, 'The Myth of Renaissance Atheism and the French Tradition of Free Thought', *Journal of the History of Philosophy*, 1968, 6, pp. 233–43.
133. Exiles from Padua 'Patavinam libertatem, quoties meminerint, suspirant', quoted in Mabilleau, op.cit., p. 33.
134. Bouwsma op.cit., pp. 194, 236–38.
135. [Fulgenzio Micanzio] *The Life of Father Paul, of the Order of the Servie. Translated out of the Italian by a Person of Quality*, London, Moseley and Marriot, 1651, p. 49; I have modernized the spelling. Wootton, op.cit., has argued that Sarpi was an atheist.
136. Ibid., p. 53.
137. Ibid., p. 124. One of Fabricius's books was dedicated to Andreas Morosini: *De Musculi Artificio; De Ossium Dearticulationibus*, of 1614.
138. Ibid., p. 145.
139. Ibid., pp. 24, 29–30.
140. Frances Yates, 'Paolo Sarpi's *History of the Council of Trent*', first published 1944, reprinted in her *Renaissance and Reform: The Italian Contribution. Collected Essays, Volume II*, London, Routledge and Kegan Paul, 1983, pp. 189–217, see esp. p. 191.

Index